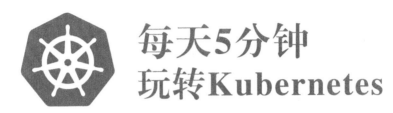

每天5分钟
玩转Kubernetes

CloudMan 著

清华大学出版社
北京

内 容 简 介

Kubernetes 是容器编排引擎的事实标准，是继大数据、云计算和 Docker 之后又一热门技术，而且未来相当一段时间内都会非常流行。对于 IT 行业来说，这是一项非常有价值的技术。对于 IT 从业者来说，掌握容器技术既是市场的需要，也是提升自我价值的重要途径。

本书共 15 章，系统介绍了 Kubernetes 的架构、重要概念、安装部署方法、运行管理应用的技术、网络存储管理、集群监控和日志管理等重要内容。书中通过大量实操案例深入浅出地讲解 Kubernetes 核心技术，是一本从入门到进阶的实用 Kubernetes 操作指导手册。读者在学习的过程中，可以跟着教程进行操作，在实践中掌握 Kubernetes 的核心技能。在之后的工作中，则可以将本教程作为参考书，按需查找相关知识点。

本书主要面向微服务软件开发人员，以及 IT 实施和运维工程师等相关人员，也适合作为高等院校和培训学校相关专业的教学参考书。

本书封面贴有清华大学出版社防伪标签，无标签者不得销售
版权所有，侵权必究。举报：010-62782989，beiqinquan@tup.tsinghua.edu.cn。

图书在版编目（CIP）数据

每天 5 分钟玩转 Kubernetes / CloudMan 著. — 北京：清华大学出版社，2018（2025.1 重印）
ISBN 978-7-302-49667-0

Ⅰ. ①每… Ⅱ. ①C… Ⅲ. ①Linux 操作系统－程序设计 Ⅳ. ①TP316.85

中国版本图书馆 CIP 数据核字(2018)第 033859 号

责任编辑：夏毓彦
封面设计：王　翔
责任校对：闫秀华
责任印制：杨　艳

出版发行：清华大学出版社
网　　址：https://www.tup.com.cn, https://www.wqxuetang.com
地　　址：北京清华大学学研大厦 A 座　　邮　编：100084
社 总 机：010-83470000　　邮　购：010-62786544
投稿与读者服务：010-62776969，c-service@tup.tsinghua.edu.cn
质量反馈：010-62772015，zhiliang@tup.tsinghua.edu.cn

印 装 者：三河市龙大印装有限公司
经　　销：全国新华书店
开　　本：190mm×260mm　　印　张：11.5　　字　数：294 千字
版　　次：2018 年 4 月第 1 版　　印　次：2025 年 1 月第 12 次印刷
定　　价：49.00 元

产品编号：079113-02

前　言

写在最前面

《每天 5 分钟玩转 Kubernetes》是一本系统学习 Kubernetes 的教程，有下面两个特点：

- 系统讲解当前最流行的容器编排引擎 Kubernetes
  包括安装部署、应用管理、网络、存储、监控、日志管理等多个方面。
- 重实践并兼顾理论
  通过大量实验和操作带领大家学习 Kubernetes。

为什么要写这个

因为 Kubernetes 非常热门，但学习门槛高。

2017 年 9 月，Mesosphere 宣布支持 Kubernetes；10 月，Docker 宣布将在新版本中加入对 Kubernetes 的原生支持。至此，容器编排引擎领域的三足鼎立时代结束，Kubernetes 赢得全面胜利。

其实早在 2015 年 5 月，Kubernetes 在 Google 上的搜索热度就已经超过了 Mesos 和 Docker Swarm，从那之后便是一路飙升，将对手"甩开了十几条街"。

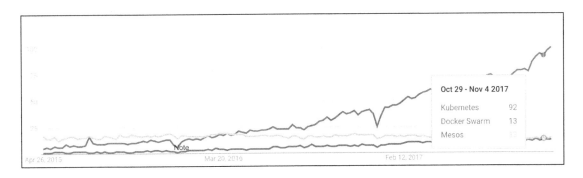

目前，AWS、Azure、Google、阿里云、腾讯云等主流公有云提供的是基于 Kubernetes 的容器服务。Rancher、CoreOS、IBM、Mirantis、Oracle、Red Hat、VMWare 等无数厂商也在大力研发和推广基于 Kubernetes 的容器 CaaS 或 PaaS 产品。可以说，Kubernetes 是当前容器行业最热门的。

每一轮新技术的兴起，无论对公司还是个人既是机会也是挑战。这项新技术未来必将成为主流，那么作为 IT 从业者，正确的做法就是尽快掌握。因为：

（1）新技术意味着新的市场和新的需求。初期掌握这种技术的人不是很多，而市场需求会越来越大，因而会形成供不应求的卖方市场，物以稀为贵，这对技术人员将是一个难得的价值提升机会。

（2）学习新技术需要时间和精力，早起步早成材。

机会讲过了，咱们再来看看挑战。

新技术往往意味着技术上的突破和创新，会有不少新的概念和方法。

对于 Kubernetes 这项平台级技术，覆盖的技术范围非常广，包括计算、网络、存储、高可用、监控、日志管理等多个方面，要掌握这些新技术对 IT 老兵尚有不小难度，更别说新人了。

写给谁看

这套教程的目标读者包括：

IT 实施和运维工程师

越来越多的应用将以容器的方式在开发、测试和生产环境中运行。掌握基于 Kubernetes 的容器平台运维能力将成为实施和运维工程师的核心竞争力。

软件开发人员

基于容器的微服务架构（Microservice Architecture）会逐渐成为开发应用系统的主流，而 Kubernetes 将是运行微服务应用的理想平台，市场将需要大量具备 Kubernetes 技能的应用程序开发人员。

我自己

CloudMan 坚信最好的学习方法是分享。编写这本教程的同时也是对自己学习和实践 Kubernetes 技术的总结。对于知识，只有把它写出来并能够让其他人理解，才能说明自己真正掌握了。

著　者

2018 年 1 月

目　录

第 1 章　先把 Kubernetes 跑起来 1

    1.1　先跑起来 1

    1.2　创建 Kubernetes 集群 2

    1.3　部署应用 4

    1.4　访问应用 5

    1.5　Scale 应用 6

    1.6　滚动更新 7

    1.7　小结 8

第 2 章　重要概念 9

第 3 章　部署 Kubernetes Cluster 13

    3.1　安装 Docker 14

    3.2　安装 kubelet、kubeadm 和 kubectl 14

    3.3　用 kubeadm 创建 Cluster 14

        3.3.1　初始化 Master 14

        3.3.2　配置 kubectl 16

        3.3.3　安装 Pod 网络 16

        3.3.4　添加 k8s-node1 和 k8s-node2 16

    3.4　小结 18

第 4 章　Kubernetes 架构 19

    4.1　Master 节点 19

    4.2　Node 节点 20

    4.3　完整的架构图 21

    4.4　用例子把它们串起来 22

    4.5　小结 24

第 5 章 运行应用 25

5.1 Deployment 25
5.1.1 运行 Deployment 25
5.1.2 命令 vs 配置文件 29
5.1.3 Deployment 配置文件简介 30
5.1.4 伸缩 31
5.1.5 Failover 33
5.1.6 用 label 控制 Pod 的位置 33

5.2 DaemonSet 36
5.2.1 kube-flannel-ds 36
5.2.2 kube-proxy 37
5.2.3 运行自己的 DaemonSet 38

5.3 Job 40
5.3.1 Pod 失败的情况 41
5.3.2 Job 的并行性 43
5.3.3 定时 Job 45

5.4 小结 48

第 6 章 通过 Service 访问 Pod 49

6.1 创建 Service 49
6.2 Cluster IP 底层实现 51
6.3 DNS 访问 Service 53
6.4 外网如何访问 Service 55
6.5 小结 58

第 7 章 Rolling Update 59

7.1 实践 59
7.2 回滚 61
7.3 小结 63

第 8 章 Health Check 64

8.1 默认的健康检查 64
8.2 Liveness 探测 65
8.3 Readiness 探测 67

8.4 Health Check 在 Scale Up 中的应用 69

8.5 Health Check 在滚动更新中的应用 71

8.6 小结 75

第 9 章 数据管理 76

9.1 Volume 76

9.1.1 emptyDir 76

9.1.2 hostPath 78

9.1.3 外部 Storage Provider 79

9.2 PersistentVolume & PersistentVolumeClaim 81

9.2.1 NFS PersistentVolume 81

9.2.2 回收 PV 84

9.2.3 PV 动态供给 86

9.3 一个数据库例子 87

9.4 小结 91

第 10 章 Secret & Configmap 92

10.1 创建 Secret 92

10.2 查看 Secret 93

10.3 在 Pod 中使用 Secret 94

10.3.1 Volume 方式 94

10.3.2 环境变量方式 96

10.4 ConfigMap 97

10.5 小结 100

第 11 章 Helm—Kubernetes 的包管理器 101

11.1 Why Helm 101

11.2 Helm 架构 103

11.3 安装 Helm 104

11.3.1 Helm 客户端 104

11.3.2 Tiller 服务器 105

11.4 使用 Helm 106

11.5 chart 详解 109

11.5.1 chart 目录结构 109

	11.5.2 chart 模板	113
	11.5.3 再次实践 MySQL chart	115
	11.5.4 升级和回滚 release	118
	11.5.5 开发自己的 chart	119
11.6	小结	126

第 12 章 网　络 127

12.1	Kubernetes 网络模型	127
12.2	各种网络方案	128
12.3	Network Policy	129
	12.3.1 部署 Canal	129
	12.3.2 实践 Network Policy	130
12.4	小结	135

第 13 章 Kubernetes Dashboard 136

13.1	安装	136
13.2	配置登录权限	137
13.3	Dashboard 界面结构	139
13.4	典型使用场景	140
	13.4.1 部署 Deployment	140
	13.4.2 在线操作	141
	13.4.3 查看资源详细信息	142
	13.4.4 查看 Pod 日志	142
13.5	小结	143

第 14 章 Kubernetes 集群监控 144

14.1	Weave Scope	144
	14.1.1 安装 Scope	144
	14.1.2 使用 Scope	145
14.2	Heapster	151
	14.2.1 部署	151
	14.2.2 使用	152
14.3	Prometheus Operator	155
	14.3.1 Prometheus 架构	159

  14.3.2 Prometheus Operator 架构 161

  14.3.3 部署 Prometheus Operator 162

 14.4 小结 167

第 15 章 Kubernetes 集群日志管理 168

 15.1 部署 168

 15.2 小结 173

写在最后 174

第 1 章
先把 Kubernetes 跑起来

Kubernetes（K8s）是 Google 在 2014 年发布的一个开源项目。

据说 Google 的数据中心里运行着 20 多亿个容器，而且 Google 十年前就开始使用容器技术。

最初，Google 开发了一个叫 Borg 的系统（现在命名为 Omega）来调度如此庞大数量的容器和工作负载。在积累了这么多年的经验后，Google 决定重写这个容器管理系统，并将其贡献到开源社区，让全世界都能受益。

这个项目就是 Kubernetes。简单地讲，Kubernetes 是 Google Omega 的开源版本。

从 2014 年第一个版本发布以来，Kubernetes 迅速获得开源社区的追捧，包括 Red Hat、VMware、Canonical 在内的很多有影响力的公司加入到开发和推广的阵营。目前 Kubernetes 已经成为发展最快、市场占有率最高的容器编排引擎产品。

Kubernetes 一直在快速地开发和迭代。本书我们将以 v1.7 和 v1.8 为基础学习 Kubernetes。我们会讨论 Kubernetes 重要的概念和架构，学习 Kubernetes 如何编排容器，包括优化资源利用、高可用、滚动更新、网络插件、服务发现、监控、数据管理、日志管理等。

下面就让我们开始 Kubernetes 的探险之旅。

1.1 先跑起来

按照一贯的学习思路，我们会在最短时间内搭建起一个可用系统，这样就能够尽快建立起对学习对象的感性认识。先把玩起来，快速了解基本概念、功能和使用场景。

越是门槛高的知识，就越需要有这么一个最小可用系统。如果直接上来就学习理论知识和概念，很容易从入门到放弃。

当然，要搭建这么一个可运行的系统通常也不会太容易，不过很幸运，Kubernetes 官网已经为大家准备好了现成的最小可用系统。

kubernetes.io 开发了一个交互式教程，通过 Web 浏览器就能使用预先部署好的一个 Kubernetes 集群，快速体验 Kubernetes 的功能和应用场景，下面我就带着大家去玩一下。

打开 https://kubernetes.io/docs/tutorials/kubernetes-basics/。

页面左边就能看到教程菜单，如图 1-1 所示。

图 1-1

教程会指引大家完成创建 Kubernetes 集群、部署应用、访问应用、扩展应用、更新应用等最常见的使用场景，迅速建立感性认识。

1.2 创建 Kubernetes 集群

点击教程菜单 1. Create a Cluster → Interactive Tutorial - Creating a Cluster，如图 1-2 所示。

图 1-2

显示操作界面，如图 1-3 所示。

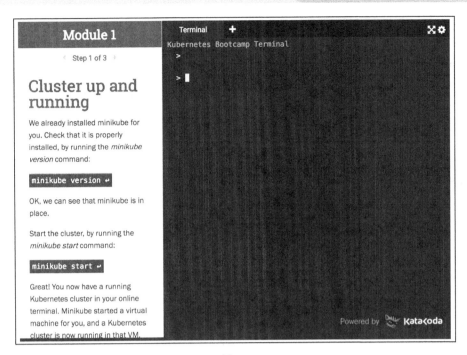

图 1-3

左边部分是操作说明。右边是 Terminal，即命令终端窗口。

按照操作说明，我们在 Terminal 中执行 minikube start，然后执行 kubectl get nodes，这样就创建好了一个单节点的 kubernetes 集群，如图 1-4 所示。

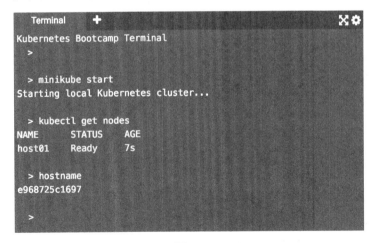

图 1-4

集群的唯一节点为 host01，需要注意的是当前执行命令的地方并不是 host01。我们是通过 Kubernetes 的命令行工具 kubectl 远程管理集群。

执行 kubectl cluster-info 查看集群信息，如图 1-5 所示。

```
> kubectl cluster-info
Kubernetes master is running at http://host01:8080
heapster is running at http://host01:8080/api/v1/proxy/namespaces/kube-system/services/heapster
kubernetes-dashboard is running at http://host01:8080/api/v1/proxy/namespaces/kube-system/services/kubernetes-dashboard
monitoring-grafana is running at http://host01:8080/api/v1/proxy/namespaces/kube-system/services/monitoring-grafana
monitoring-influxdb is running at http://host01:8080/api/v1/proxy/namespaces/kube-system/services/monitoring-influxdb

To further debug and diagnose cluster problems, use 'kubectl cluster-info dump'.
>
```

图 1-5

heapster、kubernetes-dashboard 都是集群中运行的服务。

注意：为节省篇幅，在后面的演示中，我们将简化操作步骤，详细的说明和完整步骤请参考官网在线文档。

1.3 部署应用

执行命令：

```
kubectl run kubernetes-bootcamp \
    --image=docker.io/jocatalin/kubernetes-bootcamp:v1 \
    --port=8080
```

这里我们通过 kubectl run 部署了一个应用，命名为 kubernetes-bootcamp，如图 1-6 所示。
Docker 镜像通过 --image 指定。
--port 设置应用对外服务的端口。

图 1-6

这里 Deployment 是 Kubernetes 的术语，可以理解为应用。
Kubernetes 还有一个重要术语 Pod。
Pod 是容器的集合，通常会将紧密相关的一组容器放到一个 Pod 中，同一个 Pod 中的所有容器共享 IP 地址和 Port 空间，也就是说它们在一个 network namespace 中。
Pod 是 Kubernetes 调度的最小单位，同一 Pod 中的容器始终被一起调度。
运行 kubectl get pods，查看当前的 Pod，如图 1-7 所示。

```
> kubectl get pods
NAME                                        READY    STATUS     RESTARTS   AGE
kubernetes-bootcamp-390780338-q9p1t         1/1      Running    0          11m
>
```

图 1-7

kubernetes-bootcamp-390780338-q9p1t 就是应用的 Pod。

1.4 访问应用

默认情况下，所有 Pod 只能在集群内部访问。对于上面这个例子，要访问应用只能直接访问容器的 8080 端口。为了能够从外部访问应用，我们需要将容器的 8080 端口映射到节点的端口。

执行如下命令，结果如图 1-8 所示。

```
kubectl expose deployment/kubernetes-bootcamp \
    --type="NodePort" \
    --port 8080
```

```
> kubectl expose deployment/kubernetes-bootcamp \
>     --type="NodePort" \
>     --port 8080
service "kubernetes-bootcamp" exposed
>
```

图 1-8

执行命令 kubectl get services，可以查看应用被映射到节点的哪个端口，如图 1-9 所示。

```
> kubectl get services
NAME                    CLUSTER-IP    EXTERNAL-IP    PORT(S)          AGE
kubernetes              10.0.0.1      <none>         443/TCP          2m
kubernetes-bootcamp     10.0.0.131    <nodes>        8080:32320/TCP   1m
>
```

图 1-9

这里有两个 service，可以将 service 暂时理解为端口映射，后面我们会详细讨论。

Kubernetes 是默认的 service，暂时不用考虑。kubernetes-bootcamp 是我们应用的 service，8080 端口已经映射到 host01 的 32320 端口，端口号是随机分配的，可以执行如下命令访问应用，结果如图 1-10 所示。

```
curl host01:32320
```

```
> curl host01:32320
Hello Kubernetes bootcamp! | Running on: kubernetes-bootcamp-390780338-q9p1t | v=1
>
```

图 1-10

1.5 Scale 应用

默认情况下应用只会运行一个副本，可以通过 kubectl get deployments 查看副本数，如图 1-11 所示。

```
> kubectl get deployments
NAME                    DESIRED   CURRENT   UP-TO-DATE   AVAILABLE   AGE
kubernetes-bootcamp     1         1         1            1           14m
>
```

图 1-11

执行如下命令将副本数增加到 3 个，如图 1-12 所示。

```
kubectl scale deployments/kubernetes-bootcamp --replicas=3
```

```
> kubectl scale deployments/kubernetes-bootcamp --replicas=3
deployment "kubernetes-bootcamp" scaled
>
> kubectl get deployments
NAME                    DESIRED   CURRENT   UP-TO-DATE   AVAILABLE   AGE
kubernetes-bootcamp     3         3         3            3           17m
>
```

图 1-12

通过 kubectl get pods 可以看到当前 Pod 增加到 3 个，如图 1-13 所示。

```
> kubectl get pods
NAME                                      READY   STATUS    RESTARTS   AGE
kubernetes-bootcamp-390780338-12sbg       1/1     Running   0          1m
kubernetes-bootcamp-390780338-q9p1t       1/1     Running   0          19m
kubernetes-bootcamp-390780338-swvp7       1/1     Running   0          1m
>
```

图 1-13

通过 curl 访问应用，可以看到每次请求发送到不同的 Pod，3 个副本轮询处理，这样就实现了负载均衡，如图 1-14 所示。

图 1-14

要 scale down 也很方便，执行下列命令，结果如图 1-15 所示。

```
kubectl scale deployments/kubernetes-bootcamp --replicas=2
```

图 1-15

从图 1-15 中可以看到，其中一个副本被删除了。

1.6 滚动更新

当前应用使用的 image 版本为 v1，执行如下命令将其升级到 v2，结果如图 1-16 所示。

```
kubectl set image deployments/kubernetes-bootcamp kubernetes-bootcamp=jocatalin/kubernetes-bootcamp:v2
```

图 1-16

通过 kubectl get pods 可以观察滚动更新的过程：v1 的 Pod 被逐个删除，同时启动了新的 v2 Pod。更新完成后访问新版本应用，如图 1-17 所示。

```
> curl host01:32320
Hello Kubernetes bootcamp! | Running on: kubernetes-bootcamp-2100875782-sbwkc | v=2
> curl host01:32320
Hello Kubernetes bootcamp! | Running on: kubernetes-bootcamp-2100875782-2q5k8 | v=2
>
```

图 1-17

如果要回退到 v1 版本也很容易，执行 kubectl rollout undo 命令，结果如图 1-18 所示。

```
kubectl rollout undo deployments/kubernetes-bootcamp
```

```
> kubectl rollout undo deployments/kubernetes-bootcamp
deployment "kubernetes-bootcamp" rolled back

> kubectl get pods
NAME                                    READY   STATUS              RESTARTS   AGE
kubernetes-bootcamp-2100875782-2q5k8    1/1     Running             0          7m
kubernetes-bootcamp-2100875782-sbwkc    1/1     Terminating         0          7m
kubernetes-bootcamp-390780338-9kvjj     0/1     ContainerCreating   0          5s
kubernetes-bootcamp-390780338-hmcgb     0/1     ContainerCreating   0          5s
```

图 1-18

验证版本已经回退到 v1，如图 1-19 所示。

```
> curl host01:32320
Hello Kubernetes bootcamp! | Running on: kubernetes-bootcamp-390780338-hmcgb | v=1
> curl host01:32320
Hello Kubernetes bootcamp! | Running on: kubernetes-bootcamp-390780338-9kvjj | v=1
>
```

图 1-19

1.7 小结

至此，我们已经通过官网的交互式教程快速体验了 Kubernetes 的功能和使用方法。本书的其余章节将详细讨论 Kubernetes 的架构、典型的部署方法、容器编排能力、网络方案、监控方案，帮助大家全面掌握 Kubernetes 的核心技能。

第 2 章 重要概念

在实践之前，必须先学习 Kubernetes 的几个重要概念，它们是组成 Kubernetes 集群的基石。

1. Cluster

Cluster 是计算、存储和网络资源的集合，Kubernetes 利用这些资源运行各种基于容器的应用。

2. Master

Master 是 Cluster 的大脑，它的主要职责是调度，即决定将应用放在哪里运行。Master 运行 Linux 操作系统，可以是物理机或者虚拟机。为了实现高可用，可以运行多个 Master。

3. Node

Node 的职责是运行容器应用。Node 由 Master 管理，Node 负责监控并汇报容器的状态，同时根据 Master 的要求管理容器的生命周期。Node 运行在 Linux 操作系统上，可以是物理机或者是虚拟机。

在前面交互式教程中，我们创建的 Cluster 只有一个主机 host01，它既是 Master 也是 Node，如图 2-1 所示。

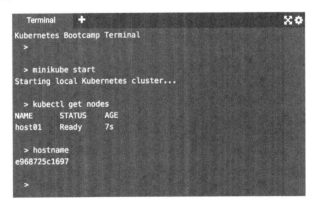

图 2-1

4. Pod

Pod 是 Kubernetes 的最小工作单元。每个 Pod 包含一个或多个容器。Pod 中的容器会作为一个整体被 Master 调度到一个 Node 上运行。

Kubernetes 引入 Pod 主要基于下面两个目的：

（1）可管理性。

有些容器天生就是需要紧密联系，一起工作。Pod 提供了比容器更高层次的抽象，将它们封装到一个部署单元中。Kubernetes 以 Pod 为最小单位进行调度、扩展、共享资源、管理生命周期。

（2）通信和资源共享。

Pod 中的所有容器使用同一个网络 namespace，即相同的 IP 地址和 Port 空间。它们可以直接用 localhost 通信。同样的，这些容器可以共享存储，当 Kubernetes 挂载 volume 到 Pod，本质上是将 volume 挂载到 Pod 中的每一个容器。

Pods 有两种使用方式：

（1）运行单一容器。

one-container-per-Pod 是 Kubernetes 最常见的模型，这种情况下，只是将单个容器简单封装成 Pod。即便是只有一个容器，Kubernetes 管理的也是 Pod 而不是直接管理容器。

（2）运行多个容器。

问题在于：哪些容器应该放到一个 Pod 中？

答案是：这些容器联系必须非常紧密，而且需要直接共享资源。

举个例子，如图 2-2 所示，这个 Pod 包含两个容器：一个是 File Puller，一个是 Web Server。

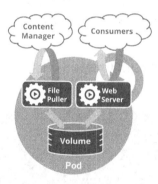

图 2-2

File Puller 会定期从外部的 Content Manager 中拉取最新的文件，将其存放在共享的 volume 中。Web Server 从 volume 读取文件，响应 Consumer 的请求。

这两个容器是紧密协作的，它们一起为 Consumer 提供最新的数据；同时它们也通过 volume 共享数据，所以放到一个 Pod 是合适的。

再来看一个反例：是否需要将 Tomcat 和 MySQL 放到一个 Pod 中？

Tomcat 从 MySQL 读取数据，它们之间需要协作，但还不至于需要放到一个 Pod 中一起部署、一起启动、一起停止。同时它们之间是通过 JDBC 交换数据，并不是直接共享存储，所以放到各自的 Pod 中更合适。

5. Controller

Kubernetes 通常不会直接创建 Pod，而是通过 Controller 来管理 Pod 的。Controller 中定义了 Pod 的部署特性，比如有几个副本、在什么样的 Node 上运行等。为了满足不同的业务场景，Kubernetes 提供了多种 Controller，包括 Deployment、ReplicaSet、DaemonSet、StatefuleSet、Job 等，我们逐一讨论。

（1）Deployment 是最常用的 Controller，比如在线教程中就是通过创建 Deployment 来部署应用的。Deployment 可以管理 Pod 的多个副本，并确保 Pod 按照期望的状态运行。

（2）ReplicaSet 实现了 Pod 的多副本管理。使用 Deployment 时会自动创建 ReplicaSet，也就是说 Deployment 是通过 ReplicaSet 来管理 Pod 的多个副本的，我们通常不需要直接使用 ReplicaSet。

（3）DaemonSet 用于每个 Node 最多只运行一个 Pod 副本的场景。正如其名称所揭示的，DaemonSet 通常用于运行 daemon。

（4）StatefuleSet 能够保证 Pod 的每个副本在整个生命周期中名称是不变的，而其他 Controller 不提供这个功能。当某个 Pod 发生故障需要删除并重新启动时，Pod 的名称会发生变化，同时 StatefuleSet 会保证副本按照固定的顺序启动、更新或者删除。

（5）Job 用于运行结束就删除的应用，而其他 Controller 中的 Pod 通常是长期持续运行。

6. Service

Deployment 可以部署多个副本，每个 Pod 都有自己的 IP，外界如何访问这些副本呢？通过 Pod 的 IP 吗？

要知道 Pod 很可能会被频繁地销毁和重启，它们的 IP 会发生变化，用 IP 来访问不太现实。

答案是 Service。

Kubernetes Service 定义了外界访问一组特定 Pod 的方式。Service 有自己的 IP 和端口，Service 为 Pod 提供了负载均衡。

Kubernetes 运行容器（Pod）与访问容器（Pod）这两项任务分别由 Controller 和 Service 执行。

7. Namespace

如果有多个用户或项目组使用同一个 Kubernetes Cluster，如何将他们创建的 Controller、Pod 等资源分开呢？

答案就是 Namespace。

Namespace 可以将一个物理的 Cluster 逻辑上划分成多个虚拟 Cluster，每个 Cluster 就是一个 Namespace。不同 Namespace 里的资源是完全隔离的。

Kubernetes 默认创建了两个 Namespace，如图 2-3 所示。

```
> kubectl get namespace
NAME          STATUS   AGE
default       Active   17s
kube-system   Active   17s
>
```

图 2-3

- default：创建资源时如果不指定，将被放到这个 Namespace 中。
- kube-system：Kubernetes 自己创建的系统资源将放到这个 Namespace 中。

第 3 章
部署 Kubernetes Cluster

本章我们将部署三个节点的 Kubernetes Cluster，如图 3-1 所示。

图 3-1

k8s-master 是 Master，k8s-node1 和 k8s-node2 是 Node。

所有节点的操作系统均为 Ubuntu 16.04，当然其他 Linux 也是可以的。

官方安装文档可以参考 https://kubernetes.io/docs/setup/independent/install-kubeadm/。

注意：Kubernetes 几乎所有的安装组件和 Docker 镜像都放在 Google 自己的网站上，这对国内的同学可能是个不小的障碍。建议是：网络障碍都必须想办法克服，不然连 Kubernetes 的门都进不了。

3.1 安装 Docker

所有节点都需要安装 Docker。

```
apt-get update && apt-get install docker.io
```

3.2 安装 kubelet、kubeadm 和 kubectl

在所有节点上安装 kubelet、kubeadm 和 kubectl。

- kubelet 运行在 Cluster 所有节点上，负责启动 Pod 和容器。
- kubeadm 用于初始化 Cluster。
- kubectl 是 Kubernetes 命令行工具。通过 kubectl 可以部署和管理应用，查看各种资源，创建、删除和更新各种组件。

```
apt-get update && apt-get install -y apt-transport-https
curl -s https://packages.cloud.google.com/apt/doc/apt-key.gpg | apt-key add -
cat <<EOF >/etc/apt/sources.list.d/kubernetes.list
deb http://apt.kubernetes.io/ kubernetes-xenial main
EOF
apt-get update
apt-get install -y kubelet kubeadm kubectl
```

3.3 用 kubeadm 创建 Cluster

完整的官方文档可以参考 https://kubernetes.io/docs/setup/independent/create-cluster-kubeadm/。

3.3.1 初始化 Master

在 Master 上执行如下命令：

```
kubeadm init --apiserver-advertise-address 192.168.56.105 --pod-network-cidr=10.244.0.0/16
```

--apiserver-advertise-address 指明用 Master 的哪个 interface 与 Cluster 的其他节点通

信。如果 Master 有多个 interface，建议明确指定，如果不指定，kubeadm 会自动选择有默认网关的 interface。

--pod-network-cidr 指定 Pod 网络的范围。Kubernetes 支持多种网络方案，而且不同网络方案对 --pod-network-cidr 有自己的要求，这里设置为 10.244.0.0/16 是因为我们将使用 flannel 网络方案，必须设置成这个 CIDR。在后面的实践中我们会切换到其他网络方案，比如 Canal。

初始化过程如图 3-2 所示。

```
root@k8s-master:~#
root@k8s-master:~# kubeadm init --apiserver-advertise-address 192.168.56.105 --pod-network-cidr=10.244.0.0/16
[kubeadm] WARNING: kubeadm is in beta, please do not use it for production clusters.
[init] Using Kubernetes version: v1.7.4
[init] Using Authorization modes: [Node RBAC]
[preflight] Running pre-flight checks                                        ①
[preflight] Starting the kubelet service
[kubeadm] WARNING: starting in 1.8, tokens expire after 24 hours by default (if you require a non-expiring token use --token-ttl 0)
[certificates] Generated CA certificate and key.
[certificates] Generated API server certificate and key.
[certificates] API Server serving cert is signed for DNS names [k8s-master kubernetes kubernetes.default kubernetes.default.svc kubernetes
96.0.1 192.168.56.105]
[certificates] Generated API server kubelet client certificate and key.
[certificates] Generated service account token signing key and public key.  ②
[certificates] Generated front-proxy CA certificate and key.
[certificates] Generated front-proxy client certificate and key.
[certificates] Valid certificates and keys now exist in "/etc/kubernetes/pki"
[kubeconfig] Wrote KubeConfig file to disk: "/etc/kubernetes/admin.conf"
[kubeconfig] Wrote KubeConfig file to disk: "/etc/kubernetes/kubelet.conf"   ③
[kubeconfig] Wrote KubeConfig file to disk: "/etc/kubernetes/controller-manager.conf"
[kubeconfig] Wrote KubeConfig file to disk: "/etc/kubernetes/scheduler.conf"
[apiclient] Created API client, waiting for the control plane to become ready ④
[apiclient] All control plane components are healthy after 26.505992 seconds
[token] Using token: d38a01.13653e584ccc1980
[apiconfig] Created RBAC rules
[addons] Applied essential addon: kube-proxy                                 ⑤
[addons] Applied essential addon: kube-dns

Your Kubernetes master has initialized successfully!                         ⑥

To start using your cluster, you need to run (as a regular user):

  mkdir -p $HOME/.kube
  sudo cp -i /etc/kubernetes/admin.conf $HOME/.kube/config                   ⑦
  sudo chown $(id -u):$(id -g) $HOME/.kube/config

You should now deploy a pod network to the cluster.
Run "kubectl apply -f [podnetwork].yaml" with one of the options listed at:  ⑧
  http://kubernetes.io/docs/admin/addons/

You can now join any number of machines by running the following on each node
as root:

  kubeadm join --token d38a01.13653e584ccc1980 192.168.56.105:6443           ⑨

root@k8s-master:~#
```

图 3-2

（1）kubeadm 执行初始化前的检查。

（2）生成 token 和证书。

（3）生成 KubeConfig 文件，kubelet 需要用这个文件与 Master 通信。

（4）安装 Master 组件，会从 Google 的 Registry 下载组件的 Docker 镜像。这一步可能会花一些时间，主要取决于网络质量。

（5）安装附加组件 kube-proxy 和 kube-dns。

（6）Kubernetes Master 初始化成功。

（7）提示如何配置 kubectl，后面会实践。

（8）提示如何安装 Pod 网络，后面会实践。

（9）提示如何注册其他节点到 Cluster，后面会实践。

3.3.2 配置 kubectl

kubectl 是管理 Kubernetes Cluster 的命令行工具，前面我们已经在所有的节点安装了 kubectl。Master 初始化完成后需要做一些配置工作，然后 kubectl 就能使用了。

依照 kubeadm init 输出的第 7 步提示，推荐用 Linux 普通用户执行 kubectl（root 会有一些问题）。

我们为用户 ubuntu 配置 kubectl：

```
su - ubuntu
mkdir -p $HOME/.kube
sudo cp -i /etc/kubernetes/admin.conf $HOME/.kube/config
sudo chown $(id -u):$(id -g) $HOME/.kube/config
```

为了使用更便捷，启用 kubectl 命令的自动补全功能：

```
echo "source <(kubectl completion bash)" >> ~/.bashrc
```

这样，用户 ubuntu 就可以使用 kubectl 了。

3.3.3 安装 Pod 网络

要让 Kubernetes Cluster 能够工作，必须安装 Pod 网络，否则 Pod 之间无法通信。

Kubernetes 支持多种网络方案，这里我们先使用 flannel，后面还会讨论 Canal。

执行如下命令部署 flannel，如图 3-3 所示。

```
kubectl apply -f https://raw.githubusercontent.com/coreos/flannel/master/Documentation/kube-flannel.yml
```

```
ubuntu@k8s-master:~$
ubuntu@k8s-master:~$ kubectl apply -f https://raw.githubusercontent.com/coreos/flannel/master/Documentation/kube-flannel.yml
clusterrole "flannel" created
clusterrolebinding "flannel" created
serviceaccount "flannel" created
configmap "kube-flannel-cfg" created
daemonset "kube-flannel-ds" created
ubuntu@k8s-master:~$
```

图 3-3

3.3.4 添加 k8s-node1 和 k8s-node2

在 k8s-node1 和 k8s-node2 上分别执行如下命令，将其注册到 Cluster 中：

```
kubeadm join --token d38a01.13653e584ccc1980 192.168.56.105:6443
```

这里的 --token 来自前面 kubeadm init 输出的第 9 步提示，如果当时没有记录下来，可以通过 kubeadm token list 查看，如图 3-4 所示。

```
root@k8s-master:~#
root@k8s-master:~# kubeadm token list
TOKEN                      TTL         EXPIRES     USAGES                   DESCRIPTION
d38a01.13653e584ccc1980    <forever>   <never>     authentication,signing   The default bootstrap token generated by 'kubeadm init'.
root@k8s-master:~#
```

图 3-4

kubeadm join 执行如图 3-5 所示。

```
root@k8s-node1:~#
root@k8s-node1:~# kubeadm join --token d38a01.13653e584ccc1980 192.168.56.105:6443
[kubeadm] WARNING: kubeadm is in beta, please do not use it for production clusters.
[preflight] Running pre-flight checks
[preflight] Starting the kubelet service
[discovery] Trying to connect to API Server "192.168.56.105:6443"
[discovery] Created cluster-info discovery client, requesting info from "https://192.168.56.105:6443"
[discovery] Cluster info signature and contents are valid, will use API Server "https://192.168.56.105:6443"
[discovery] Successfully established connection with API Server "192.168.56.105:6443"
[bootstrap] Detected server version: v1.7.4
[bootstrap] The server supports the Certificates API (certificates.k8s.io/v1beta1)
[csr] Created API client to obtain unique certificate for this node, generating keys and certificate signing request
[csr] Received signed certificate from the API server, generating KubeConfig...
[kubeconfig] Wrote KubeConfig file to disk: "/etc/kubernetes/kubelet.conf"

Node join complete:
* Certificate signing request sent to master and response
  received.
* Kubelet informed of new secure connection details.

Run 'kubectl get nodes' on the master to see this machine join.
root@k8s-node1:~#
```

图 3-5

根据提示，我们可以通过 kubectl get nodes 查看节点的状态，如图 3-6 所示。

```
ubuntu@k8s-master:~$ kubectl get nodes
NAME          STATUS      AGE     VERSION
k8s-master    NotReady    45m     v1.7.4
k8s-node1     NotReady    59s     v1.7.4
k8s-node2     NotReady    6s      v1.7.4
```

图 3-6

目前所有节点都是 NotReady，这是因为每个节点都需要启动若干组件，这些组件都是在 Pod 中运行，需要首先从 Google 下载镜像。我们可以通过如下命令查看 Pod 的状态，如图 3-7 所示。

```
kubectl get pod --all-namespaces
```

```
ubuntu@k8s-master:~$ kubectl get pod --all-namespaces
NAMESPACE     NAME                                    READY   STATUS              RESTARTS   AGE
kube-system   etcd-k8s-master                         1/1     Running             0          44m
kube-system   kube-apiserver-k8s-master               1/1     Running             0          44m
kube-system   kube-controller-manager-k8s-master      1/1     Running             0          44m
kube-system   kube-dns-2425271678-1z3pv               0/3     Pending             0          49m
kube-system   kube-flannel-ds-cqbpb                   0/2     ContainerCreating   0          6m
kube-system   kube-flannel-ds-v0p3x                   0/2     ImagePullBackOff    0          9m
kube-system   kube-flannel-ds-xk49w                   0/2     ContainerCreating   0          5m
kube-system   kube-proxy-16mg9                        0/1     ContainerCreating   0          6m
kube-system   kube-proxy-wc4j0                        1/1     Running             0          49m
kube-system   kube-proxy-xl5gd                        0/1     ContainerCreating   0          5m
kube-system   kube-scheduler-k8s-master               1/1     Running             0          44m
ubuntu@k8s-master:~$
```

图 3-7

Pending、ContainerCreating、ImagePullBackOff 都表明 Pod 没有就绪，Running 才是就

绪状态。我们可以通过 kubectl describe pod <Pod Name> 查看 Pod 的具体情况，比如：

```
kubectl describe pod kube-flannel-ds-v0p3x --namespace=kube-system
```

结果如图 3-8 所示。

图 3-8

为了节省篇幅，这里只截取命令输出的最后部分，可以看到在下载 image 时失败，如果网络质量不好，这种情况是很常见的。我们可以耐心等待，因为 Kubernetes 会重试，我们也可以自己手动执行 docker pull 去下载这个镜像。

等待一段时间，image 成功下载后，所有 Pod 都会处于 Running 状态，如图 3-9 所示。

图 3-9

这时，所有的节点都已经准备好了，Kubernetes Cluster 创建成功，如图 3-10 所示。

图 3-10

3.4 小结

本章通过 kubeadm 部署了三节点的 Kubernetes 集群，后面章节我们将在这个实验环境中学习 Kubernetes 的各项技术。

第 4 章 Kubernetes 架构

Kubernetes Cluster 由 Master 和 Node 组成，节点上运行着若干 Kubernetes 服务。

4.1 Master 节点

Master 是 Kubernetes Cluster 的大脑，运行着的 Daemon 服务包括 kube-apiserver、kube-scheduler、kube-controller-manager、etcd 和 Pod 网络（例如 flannel），如图 4-1 所示。

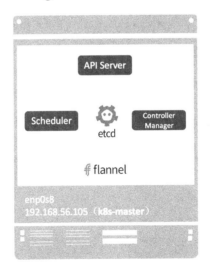

图 4-1

1. API Server（kube-apiserver）

API Server 提供 HTTP/HTTPS RESTful API，即 Kubernetes API。API Server 是 Kubernetes Cluster 的前端接口，各种客户端工具（CLI 或 UI）以及 Kubernetes 其他组件可以通过它管理 Cluster 的各种资源。

2. Scheduler（kube-scheduler）

Scheduler 负责决定将 Pod 放在哪个 Node 上运行。Scheduler 在调度时会充分考虑 Cluster 的拓扑结构，当前各个节点的负载，以及应用对高可用、性能、数据亲和性的需求。

3. Controller Manager（kube-controller-manager）

Controller Manager 负责管理 Cluster 各种资源，保证资源处于预期的状态。Controller Manager 由多种 controller 组成，包括 replication controller、endpoints controller、namespace controller、serviceaccounts controller 等。

不同的 controller 管理不同的资源。例如，replication controller 管理 Deployment、StatefulSet、DaemonSet 的生命周期，namespace controller 管理 Namespace 资源。

4. etcd

etcd 负责保存 Kubernetes Cluster 的配置信息和各种资源的状态信息。当数据发生变化时，etcd 会快速地通知 Kubernetes 相关组件。

5. Pod 网络

Pod 要能够相互通信，Kubernetes Cluster 必须部署 Pod 网络，flannel 是其中一个可选方案。

以上是 Master 上运行的组件，下面我们接着讨论 Node。

4.2 Node 节点

Node 是 Pod 运行的地方，Kubernetes 支持 Docker、rkt 等容器 Runtime。Node 上运行的 Kubernetes 组件有 kubelet、kube-proxy 和 Pod 网络（例如 flannel），如图 4-2 所示。

图 4-2

1. kubelet

kubelet 是 Node 的 agent，当 Scheduler 确定在某个 Node 上运行 Pod 后，会将 Pod

的具体配置信息（image、volume 等）发送给该节点的 kubelet，kubelet 根据这些信息创建和运行容器，并向 Master 报告运行状态。

2. kube-proxy

service 在逻辑上代表了后端的多个 Pod，外界通过 service 访问 Pod。service 接收到的请求是如何转发到 Pod 的呢？这就是 kube-proxy 要完成的工作。

每个 Node 都会运行 kube-proxy 服务，它负责将访问 service 的 TCP/UPD 数据流转发到后端的容器。如果有多个副本，kube-proxy 会实现负载均衡。

3. Pod 网络

Pod 要能够相互通信，Kubernetes Cluster 必须部署 Pod 网络，flannel 是其中一个可选方案。

4.3 完整的架构图

结合实验环境，我们得到了如图 4-3 所示的架构图。

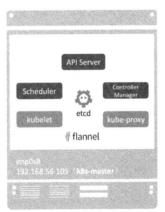

图 4-3

你可能会问：为什么 k8s-master 上也有 kubelet 和 kube-proxy 呢？

这是因为 Master 上也可以运行应用，即 Master 同时也是一个 Node。

几乎所有的 Kubernetes 组件本身也运行在 Pod 里，执行如下命令，结果如图 4-4 所示。

```
kubectl get pod --all-namespaces -o wide
```

```
ubuntu@k8s-master:~$
ubuntu@k8s-master:~$ kubectl get pod --all-namespaces -o wide
NAMESPACE     NAME                                   READY     STATUS    RESTARTS   AGE       IP               NODE
kube-system   etcd-k8s-master                        1/1       Running   0          1d        192.168.56.105   k8s-master
kube-system   kube-apiserver-k8s-master              1/1       Running   0          1d        192.168.56.105   k8s-master
kube-system   kube-controller-manager-k8s-master     1/1       Running   0          1d        192.168.56.105   k8s-master
kube-system   kube-dns-2425271678-1z3pv              3/3       Running   0          1d        10.244.1.57      k8s-node1
kube-system   kube-flannel-ds-cqbpb                  2/2       Running   4          1d        192.168.56.106   k8s-node1
kube-system   kube-flannel-ds-v0p3x                  2/2       Running   0          1d        192.168.56.105   k8s-master
kube-system   kube-flannel-ds-xk49w                  2/2       Running   0          1d        192.168.56.107   k8s-node2
kube-system   kube-proxy-16mg9                       1/1       Running   0          1d        192.168.56.106   k8s-node1
kube-system   kube-proxy-wc4j0                       1/1       Running   0          1d        192.168.56.105   k8s-master
kube-system   kube-proxy-xl5gd                       1/1       Running   0          1d        192.168.56.107   k8s-node2
kube-system   kube-scheduler-k8s-master              1/1       Running   0          1d        192.168.56.105   k8s-master
ubuntu@k8s-master:~$
```

图 4-4

Kubernetes 的系统组件都被放到 kube-system namespace 中。这里有一个 kube-dns 组件，它为 Cluster 提供 DNS 服务，我们后面会讨论到。kube-dns 是在执行 kubeadm init 时（第 5 步）作为附加组件安装的。

kubelet 是唯一没有以容器形式运行的 Kubernetes 组件，它在 Ubuntu 中通过 Systemd 服务运行，如图 4-5 所示。

```
ubuntu@k8s-master:~$
ubuntu@k8s-master:~$ sudo systemctl status kubelet.service
● kubelet.service - kubelet: The Kubernetes Node Agent
   Loaded: loaded (/lib/systemd/system/kubelet.service; enabled; vendor preset: enabled)
  Drop-In: /etc/systemd/system/kubelet.service.d
           └─10-kubeadm.conf
   Active: active (running) since Wed 2017-08-23 11:01:08 HKT; 1 day 5h ago
     Docs: http://kubernetes.io/docs/
 Main PID: 3946 (kubelet)
    Tasks: 19
   Memory: 46.4M
      CPU: 20min 29.149s
   CGroup: /system.slice/kubelet.service
           ├─3946 /usr/bin/kubelet --kubeconfig=/etc/kubernetes/kubelet.conf --require-kubeconfig=true
           └─3974 journalctl -k -f
```

图 4-5

4.4 用例子把它们串起来

为了帮助大家更好地理解 Kubernetes 架构，我们部署一个应用来演示各个组件之间是如何协作的。

执行下列命令，结果如图 4-6 所示。

```
kubectl run httpd-app --image=httpd --replicas=2
```

```
ubuntu@k8s-master:~$
ubuntu@k8s-master:~$ kubectl run httpd-app --image=httpd --replicas=2
deployment "httpd-app" created
ubuntu@k8s-master:~$
```

图 4-6

等待一段时间，应用部署完成，如图 4-7 所示。

```
ubuntu@k8s-master:~$
ubuntu@k8s-master:~$ kubectl get deployment
NAME        DESIRED   CURRENT   UP-TO-DATE   AVAILABLE   AGE
httpd-app   2         2         2            2           2m
ubuntu@k8s-master:~$
ubuntu@k8s-master:~$ kubectl get pod -o wide
NAME                           READY   STATUS    RESTARTS   AGE   IP           NODE
httpd-app-3211369089-9bgrz     1/1     Running   0          2m    10.244.1.58  k8s-node1
httpd-app-3211369089-gn6z5     1/1     Running   0          2m    10.244.2.39  k8s-node2
ubuntu@k8s-master:~$
```

图 4-7

Kubernetes 部署了 deployment httpd-app，有两个副本 Pod，分别运行在 k8s-node1 和 k8s-node2。

详细讨论整个部署过程，如图 4-8 所示。

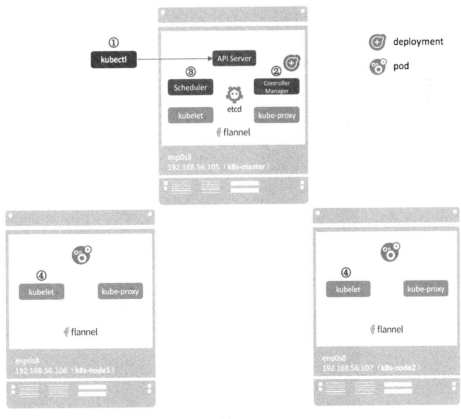

图 4-8

① kubectl 发送部署请求到 API Server。
② API Server 通知 Controller Manager 创建一个 deployment 资源。
③ Scheduler 执行调度任务，将两个副本 Pod 分发到 k8s-node1 和 k8s-node2。
④ k8s-node1 和 k8s-node2 上的 kubectl 在各自的节点上创建并运行 Pod。

补充两点：

（1）应用的配置和当前状态信息保存在 etcd 中，执行 kubectl get pod 时 API Server 会从 etcd 中读取这些数据。

（2）flannel 会为每个 Pod 都分配 IP。因为没有创建 service，所以目前 kube-proxy 还没参与进来。

4.5 小结

本章我们学习了 Kubernetes 的架构，讨论了 Master 和 Node 是哪个运行的组件和服务，并通过一个部署案例加深了对架构的理解。

第 5 章 运行应用

从本章开始，我们将通过实践深入学习 Kubernetes 的各种特性。作为容器编排引擎，最重要也是最基本的功能当然是运行容器化应用，这就是本章的内容。

5.1 Deployment

前面我们已经了解到，Kubernetes 通过各种 Controller 来管理 Pod 的生命周期。为了满足不同业务场景，Kubernetes 开发了 Deployment、ReplicaSet、DaemonSet、StatefuleSet、Job 等多种 Controller。我们首先学习最常用的 Deployment。

5.1.1 运行 Deployment

先从例子开始，运行一个 Deployment：

```
kubectl run nginx-deployment --image=nginx:1.7.9 --replicas=2
```

上面的命令将部署包含两个副本的 Deployment nginx-deployment，容器的 image 为 nginx:1.7.9。

下面详细分析 Kubernetes 都做了些什么工作，如图 5-1 所示。

```
ubuntu@k8s-master:~$
ubuntu@k8s-master:~$ kubectl run nginx-deployment --image=nginx:1.7.9 --replicas=2
deployment "nginx-deployment" created
ubuntu@k8s-master:~$
ubuntu@k8s-master:~$ kubectl get deployment nginx-deployment
NAME               DESIRED   CURRENT   UP-TO-DATE   AVAILABLE   AGE
nginx-deployment   2         2         2            2           23s
ubuntu@k8s-master:~$
```

图 5-1

在图 5-1 中，通过 kubectl get deployment 命令查看 nginx-deployment 的状态，输出显示两个副本正常运行。

接下来我们用 kubectl describe deployment 了解更详细的信息，如图 5-2 和图 5-3 所示。

```
ubuntu@k8s-master:~$
ubuntu@k8s-master:~$ kubectl describe deployment nginx-deployment
Name:                   nginx-deployment
Namespace:              default
CreationTimestamp:      Mon, 28 Aug 2017 10:28:32 +0800
Labels:                 run=nginx-deployment
Annotations:            deployment.kubernetes.io/revision=1
Selector:               run=nginx-deployment
Replicas:               2 desired | 2 updated | 2 total | 2 available | 0 unavailable
StrategyType:           RollingUpdate
MinReadySeconds:        0
RollingUpdateStrategy:  1 max unavailable, 1 max surge
Pod Template:
  Labels:       run=nginx-deployment
  Containers:
   nginx-deployment:
    Image:       nginx:1.7.9
    Port:        <none>
    Environment: <none>
    Mounts:      <none>
  Volumes:       <none>
Conditions:
  Type          Status  Reason
  ----          ------  ------
  Available     True    MinimumReplicasAvailable
```

图 5-2

```
Type     Reason            Message
------   ------            -------
Normal   ScalingReplicaSet Scaled up replica set nginx-deployment-1260880958 to 2
```

图 5-3

大部分内容都是自解释的,我们重点看图 5-3。这里告诉我们创建了一个 ReplicaSet nginx-deployment-1260880958,Events 是 Deployment 的日志,记录了 ReplicaSet 的启动过程。通过上面的分析,也验证了 Deployment 通过 ReplicaSet 来管理 Pod 的事实。接着我们将注意力切换到 nginx-deployment-1260880958,执行 kubectl describe replicaset,如图 5-4 所示。

图 5-4

两个副本已经就绪,用 kubectl describe replicaset 查看详细信息,如图 5-5 和图 5-6 所示。

```
ubuntu@k8s-master:~$
ubuntu@k8s-master:~$ kubectl describe replicaset nginx-deployment-1260880958
Name:           nginx-deployment-1260880958
Namespace:      default
Selector:       pod-template-hash=1260880958,run=nginx-deployment
Labels:         pod-template-hash=1260880958
                run=nginx-deployment
Annotations:    deployment.kubernetes.io/desired-replicas=2
                deployment.kubernetes.io/max-replicas=3
                deployment.kubernetes.io/revision=1
Controlled By:  Deployment/nginx-deployment
Replicas:       2 current / 2 desired
Pods Status:    2 Running / 0 Waiting / 0 Succeeded / 0 Failed
Pod Template:
  Labels:       pod-template-hash=1260880958
                run=nginx-deployment
  Containers:
   nginx-deployment:
    Image:        nginx:1.7.9
    Port:         <none>
    Environment:  <none>
    Mounts:       <none>
```

图 5-5

```
Type     Reason            Message
----     ------            -------
Normal   SuccessfulCreate  Created pod: nginx-deployment-1260880958-kn8w3
Normal   SuccessfulCreate  Created pod: nginx-deployment-1260880958-rpjdc
```

图 5-6

Controlled By 指明此 ReplicaSet 是由 Deployment nginx-deployment 创建的。图 5-6 是两个副本 Pod 创建的日志。接着我们来看 Pod，执行 kubectl get pod，如图 5-7 所示。

```
ubuntu@k8s-master:~$
ubuntu@k8s-master:~$ kubectl get pod
NAME                                    READY   STATUS    RESTARTS   AGE
nginx-deployment-1260880958-kn8w3       1/1     Running   0          7m
nginx-deployment-1260880958-rpjdc       1/1     Running   0          7m
ubuntu@k8s-master:~$
```

图 5-7

两个副本 Pod 都处于 Running 状态，然后用 kubectl describe pod 查看更详细的信息，如图 5-8 和图 5-9 所示。

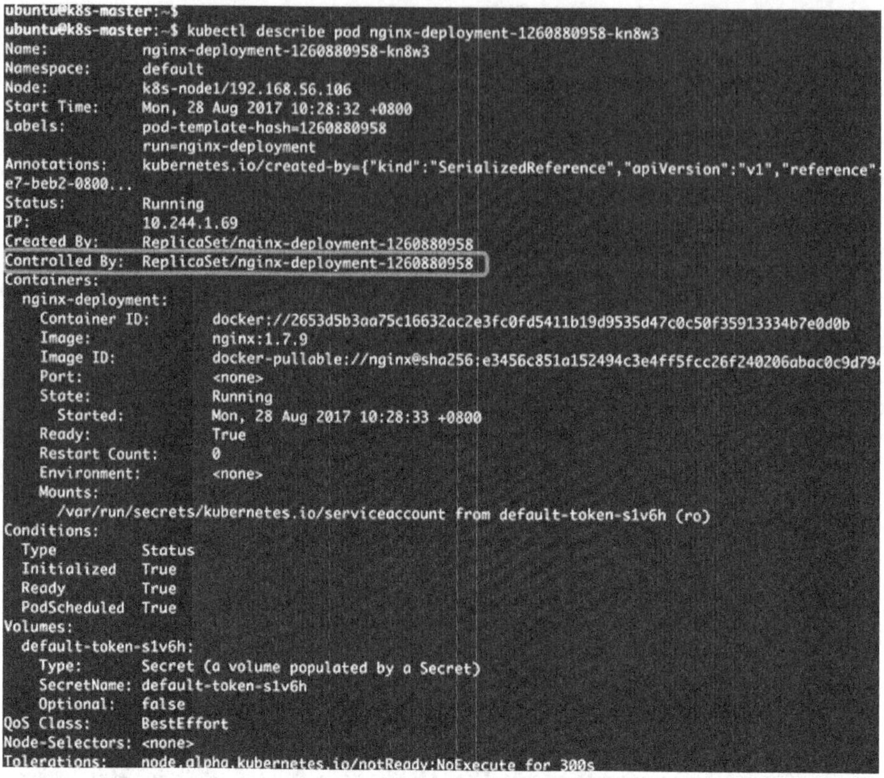

```
ubuntu@k8s-master:~$
ubuntu@k8s-master:~$ kubectl describe pod nginx-deployment-1260880958-kn8w3
Name:           nginx-deployment-1260880958-kn8w3
Namespace:      default
Node:           k8s-node1/192.168.56.106
Start Time:     Mon, 28 Aug 2017 10:28:32 +0800
Labels:         pod-template-hash=1260880958
                run=nginx-deployment
Annotations:    kubernetes.io/created-by={"kind":"SerializedReference","apiVersion":"v1","reference"
e7-beb2-0800...
Status:         Running
IP:             10.244.1.69
Created By:     ReplicaSet/nginx-deployment-1260880958
Controlled By:  ReplicaSet/nginx-deployment-1260880958
Containers:
  nginx-deployment:
    Container ID:       docker://2653d5b3aa75c16632ac2e3fc0fd5411b19d9535d47c0c50f35913334b7e0d0b
    Image:              nginx:1.7.9
    Image ID:           docker-pullable://nginx@sha256:e3456c851a152494c3e4ff5fcc26f240206abac0c9d79
    Port:               <none>
    State:              Running
      Started:          Mon, 28 Aug 2017 10:28:33 +0800
    Ready:              True
    Restart Count:      0
    Environment:        <none>
    Mounts:
      /var/run/secrets/kubernetes.io/serviceaccount from default-token-s1v6h (ro)
Conditions:
  Type          Status
  Initialized   True
  Ready         True
  PodScheduled  True
Volumes:
  default-token-s1v6h:
    Type:       Secret (a volume populated by a Secret)
    SecretName: default-token-s1v6h
    Optional:   false
QoS Class:      BestEffort
Node-Selectors: <none>
Tolerations:    node.alpha.kubernetes.io/notReady:NoExecute for 300s
```

图 5-8

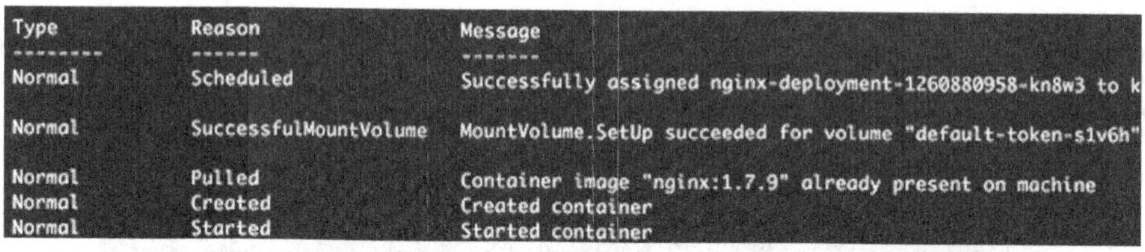

```
Type      Reason                 Message
--------  ------                 -------
Normal    Scheduled              Successfully assigned nginx-deployment-1260880958-kn8w3 to k
Normal    SuccessfulMountVolume  MountVolume.SetUp succeeded for volume "default-token-s1v6h"
Normal    Pulled                 Container image "nginx:1.7.9" already present on machine
Normal    Created                Created container
Normal    Started                Started container
```

图 5-9

Controlled By 指明此 Pod 是由 ReplicaSet nginx-deployment-1260880958 创建的。Events 记录了 Pod 的启动过程。如果操作失败（比如 image 不存在），也能在这里查到原因。

总结一下这个过程中，如图 5-10 所示。

（1）用户通过 kubectl 创建 Deployment。

（2）Deployment 创建 ReplicaSet。

（3）ReplicaSet 创建 Pod。

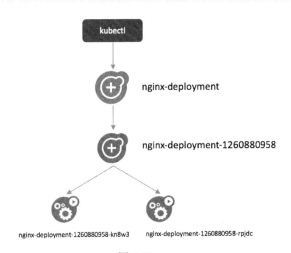

图 5-10

从图 5-10 也可以看出，对象的命名方式是"子对象的名字 = 父对象名字 + 随机字符串或数字"。

5.1.2 命令 vs 配置文件

Kubernetes 支持两种创建资源的方式：

（1）用 kubectl 命令直接创建，比如 "kubectl run nginx-deployment --image=nginx:1.7.9 --replicas=2"，在命令行中通过参数指定资源的属性。

（2）通过配置文件和 kubectl apply 创建。要完成前面同样的工作，可执行命令 "kubectl apply -f nginx.yml"，nginx.yml 的内容如图 5-11 所示。

图 5-11

资源的属性写在配置文件中，文件格式为 YAML。

下面对这两种方式进行比较。

（1）基于命令的方式：

- 简单、直观、快捷，上手快。
- 适合临时测试或实验。

（2）基于配置文件的方式：

- 配置文件描述了 What，即应用最终要达到的状态。
- 配置文件提供了创建资源的模板，能够重复部署。
- 可以像管理代码一样管理部署。
- 适合正式的、跨环境的、规模化部署。
- 这种方式要求熟悉配置文件的语法，有一定难度。

后面我们都将采用配置文件的方式，大家需要尽快熟悉和掌握。

kubectl apply 不但能够创建 Kubernetes 资源，也能对资源进行更新，非常方便。不过 Kubernets 还提供了几个类似的命令，例如 kubectl create、kubectl replace、kubectl edit 和 kubectl patch。

为避免造成不必要的困扰，我们会尽量只使用 kubectl apply，此命令已经能够应对百分之九十多的场景，事半功倍。

5.1.3 Deployment 配置文件简介

既然要用 YAML 配置文件部署应用，现在就很有必要了解一下 Deployment 的配置格式了，其他 Controller（比如 DaemonSet）非常类似。

以 nginx-deployment 为例，配置文件如图 5-12 所示。

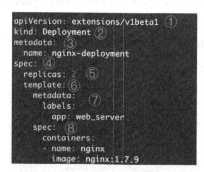

图 5-12

① apiVersion 是当前配置格式的版本。
② kind 是要创建的资源类型，这里是 Deployment。
③ metadata 是该资源的元数据，name 是必需的元数据项。
④ spec 部分是该 Deployment 的规格说明。
⑤ replicas 指明副本数量，默认为 1。
⑥ template 定义 Pod 的模板，这是配置文件的重要部分。
⑦ metadata 定义 Pod 的元数据，至少要定义一个 label。label 的 key 和 value 可以任意指定。
⑧ spec 描述 Pod 的规格，此部分定义 Pod 中每一个容器的属性，name 和 image 是必需的。

此 nginx.yml 是一个最简单的 Deployment 配置文件，后面我们学习 Kubernetes 各项功能时会逐步丰富这个文件。

执行 kubectl apply -f nginx.yml，如图 5-13 所示。

```
ubuntu@k8s-master:~$
ubuntu@k8s-master:~$ kubectl apply -f nginx.yml
deployment "nginx-deployment" created
ubuntu@k8s-master:~$
```

图 5-13

部署成功。同样，也可以通过 kubectl get 查看 nginx-deployment 的各种资源，如图 5-14 所示。

```
ubuntu@k8s-master:~$
ubuntu@k8s-master:~$ kubectl get deployment
NAME               DESIRED   CURRENT   UP-TO-DATE   AVAILABLE   AGE
nginx-deployment   2         2         2            2           17s
ubuntu@k8s-master:~$ kubectl get replicaset
NAME                          DESIRED   CURRENT   READY   AGE
nginx-deployment-2721169382   2         2         2       29s
ubuntu@k8s-master:~$ kubectl get pod -o wide
NAME                                READY   STATUS    RESTARTS   AGE   IP            NODE
nginx-deployment-2721169382-bl3cn   1/1     Running   0          41s   10.244.2.90   k8s-node2
nginx-deployment-2721169382-x6f3z   1/1     Running   0          41s   10.244.1.80   k8s-node1
ubuntu@k8s-master:~$
```

图 5-14

Deployment、ReplicaSet、Pod 都已经就绪。如果要删除这些资源，执行 kubectl delete deployment nginx-deployment 或者 kubectl delete -f nginx.yml，如图 5-15 所示。

```
ubuntu@k8s-master:~$
ubuntu@k8s-master:~$ kubectl delete -f nginx.yml
deployment "nginx-deployment" deleted
ubuntu@k8s-master:~$
```

图 5-15

5.1.4 伸缩

伸缩是指在线增加或减少 Pod 的副本数。

Deployment nginx-deployment 初始是两个副本，如图 5-16 所示。

```
ubuntu@k8s-master:~$
ubuntu@k8s-master:~$ kubectl apply -f nginx.yml
deployment "nginx-deployment" created
ubuntu@k8s-master:~$
ubuntu@k8s-master:~$ kubectl get pod -o wide
NAME                                READY   STATUS    RESTARTS   AGE   IP            NODE
nginx-deployment-2721169382-lgzzf   1/1     Running   0          8s    10.244.2.91   k8s-node2
nginx-deployment-2721169382-pt53w   1/1     Running   0          8s    10.244.1.81   k8s-node1
ubuntu@k8s-master:~$
```

图 5-16

k8s-node1 和 k8s-node2 上各跑了一个副本。现在修改 nginx.yml 文件，将副本改成 5 个，如图 5-17 所示。

```
apiVersion: extensions/v1beta1
kind: Deployment
metadata:
  name: nginx-deployment
spec:
  replicas: 5
  template:
    metadata:
      labels:
        app: web_server
    spec:
      containers:
      - name: nginx
        image: nginx:1.7.9
```

图 5-17

再次执行 kubectl apply，如图 5-18 所示。

```
ubuntu@k8s-master:~$
ubuntu@k8s-master:~$ kubectl apply -f nginx.yml
deployment "nginx-deployment" configured
ubuntu@k8s-master:~$
ubuntu@k8s-master:~$ kubectl get pod -o wide
NAME                                READY   STATUS    RESTARTS   AGE   IP            NODE
nginx-deployment-2721169382-gbjc2   1/1     Running   0          8s    10.244.2.93   k8s-node2
nginx-deployment-2721169382-lgzzf   1/1     Running   0          4m    10.244.2.91   k8s-node2
nginx-deployment-2721169382-mbrln   1/1     Running   0          8s    10.244.2.92   k8s-node2
nginx-deployment-2721169382-pt53w   1/1     Running   0          4m    10.244.1.81   k8s-node1
nginx-deployment-2721169382-s6hlx   1/1     Running   0          8s    10.244.1.82   k8s-node1
ubuntu@k8s-master:~$
```

图 5-18

三个新副本被创建并调度到 k8s-node1 和 k8s-node2 上。

出于安全考虑，默认配置下 Kubernetes 不会将 Pod 调度到 Master 节点。如果希望将 k8s-master 也当作 Node 使用，可以执行如下命令：

kubectl taint node k8s-master node-role.kubernetes.io/master-

如果要恢复 Master Only 状态，执行如下命令：

kubectl taint node k8s-master node-role.kubernetes.io/master="":NoSchedule

接下来修改配置文件，将副本数减少为 3 个，重新执行 kubectl apply，如图 5-19 所示。

```
ubuntu@k8s-master:~$
ubuntu@k8s-master:~$ kubectl apply -f nginx.yml
deployment "nginx-deployment" configured
ubuntu@k8s-master:~$
ubuntu@k8s-master:~$ kubectl get pod -o wide
NAME                                READY   STATUS        RESTARTS   AGE   IP            NODE
nginx-deployment-2721169382-gbjc2   1/1     Running       0          14m   10.244.2.93   k8s-node2
nginx-deployment-2721169382-lgzzf   1/1     Running       0          18m   10.244.2.91   k8s-node2
nginx-deployment-2721169382-mbrln   0/1     Terminating   0          14m   <none>        k8s-node2
nginx-deployment-2721169382-pt53w   1/1     Running       0          18m   10.244.1.81   k8s-node1
ubuntu@k8s-master:~$ kubectl get pod -o wide
NAME                                READY   STATUS    RESTARTS   AGE   IP            NODE
nginx-deployment-2721169382-gbjc2   1/1     Running   0          14m   10.244.2.93   k8s-node2
nginx-deployment-2721169382-lgzzf   1/1     Running   0          18m   10.244.2.91   k8s-node2
nginx-deployment-2721169382-pt53w   1/1     Running   0          18m   10.244.1.81   k8s-node1
ubuntu@k8s-master:~$
```

图 5-19

可以看到两个副本被删除，最终保留了 3 个副本。

5.1.5 Failover

下面我们模拟 k8s-node2 故障，关闭该节点，如图 5-20 所示。

图 5-20

等待一段时间，Kubernetes 会检查到 k8s-node2 不可用，将 k8s-node2 上的 Pod 标记为 Unknown 状态，并在 k8s-node1 上新创建两个 Pod，维持总副本数为 3，如图 5-21 所示。

图 5-21

当 k8s-node2 恢复后，Unknown 的 Pod 会被删除，不过已经运行的 Pod 不会重新调度回 k8s-node2，如图 5-22 所示。

图 5-22

删除 nginx-deployment，如图 5-23 所示。

图 5-23

5.1.6 用 label 控制 Pod 的位置

默认配置下，Scheduler 会将 Pod 调度到所有可用的 Node。不过有些情况我们希望将

Pod 部署到指定的 Node，比如将有大量磁盘 I/O 的 Pod 部署到配置了 SSD 的 Node；或者 Pod 需要 GPU，需要运行在配置了 GPU 的节点上。

Kubernetes 是通过 label 来实现这个功能的。

label 是 key-value 对，各种资源都可以设置 label，灵活添加各种自定义属性。比如执行如下命令标注 k8s-node1 是配置了 SSD 的节点。

```
kubectl label node k8s-node1 disktype=ssd
```

然后通过 kubectl get node --show-labels 查看节点的 label，如图 5-24 所示。

图 5-24

disktype=ssd 已经成功添加到 k8s-node1，除了 disktype，Node 还有几个 Kubernetes 自己维护的 label。

有了 disktype 这个自定义 label，接下来就可以指定将 Pod 部署到 k8s-node1。编辑 nginx.yml，如图 5-25 所示。

```
apiVersion: apps/v1beta1
kind: Deployment
metadata:
  name: nginx-deployment
spec:
  replicas: 6
  template:
    metadata:
      labels:
        app: web_server
    spec:
      containers:
      - name: nginx
        image: nginx:1.7.9
      nodeSelector:
        disktype: ssd
```

图 5-25

在 Pod 模板的 spec 里通过 nodeSelector 指定将此 Pod 部署到具有 label disktype=ssd 的 Node 上。

部署 Deployment 并查看 Pod 的运行节点，如图 5-26 所示。

```
ubuntu@k8s-master:~$
ubuntu@k8s-master:~$ kubectl apply -f nginx.yml
deployment "nginx-deployment" created
ubuntu@k8s-master:~$ kubectl get pod -o wide
NAME                                READY   STATUS    RESTARTS   AGE   IP             NODE
nginx-deployment-204403116-1kk23    1/1     Running   0          23s   10.244.1.97    k8s-node1
nginx-deployment-204403116-gcnkz    1/1     Running   0          23s   10.244.1.101   k8s-node1
nginx-deployment-204403116-kmbwr    1/1     Running   0          23s   10.244.1.96    k8s-node1
nginx-deployment-204403116-kzpnf    1/1     Running   0          23s   10.244.1.100   k8s-node1
nginx-deployment-204403116-vbz47    1/1     Running   0          23s   10.244.1.98    k8s-node1
nginx-deployment-204403116-vvh54    1/1     Running   0          23s   10.244.1.99    k8s-node1
ubuntu@k8s-master:~$
```

图 5-26

全部 6 个副本都运行在 k8s-node1 上，符合我们的预期。

要删除 label disktype，执行如下命令：

```
kubectl label node k8s-node1 disktype-
```

- 即删除，如图 5-27 所示。

```
ubuntu@k8s-master:~$
ubuntu@k8s-master:~$ kubectl label node k8s-node1 disktype-
node "k8s-node1" labeled
ubuntu@k8s-master:~$ kubectl get node --show-labels
NAME         STATUS   AGE   VERSION   LABELS
k8s-master   Ready    8d    v1.7.4    beta.kubernetes.io/arch=amd64,beta.kubernetes.io/os=linux,kubernetes.io/hostname=k8s-master,node-role.kubernetes.io/master=
k8s-node1    Ready    7d    v1.7.4    beta.kubernetes.io/arch=amd64,beta.kubernetes.io/os=linux,kubernetes.io/hostname=k8s-node1
k8s-node2    Ready    7d    v1.7.4    beta.kubernetes.io/arch=amd64,beta.kubernetes.io/os=linux,kubernetes.io/hostname=k8s-node2
ubuntu@k8s-master:~$
```

图 5-27

不过此时 Pod 并不会重新部署，依然在 k8s-node1 上运行，如图 5-28 所示。

```
ubuntu@k8s-master:~$
ubuntu@k8s-master:~$ kubectl get pod -o wide
NAME                                READY   STATUS    RESTARTS   AGE   IP             NODE
nginx-deployment-204403116-1kk23    1/1     Running   0          6m    10.244.1.97    k8s-node1
nginx-deployment-204403116-gcnkz    1/1     Running   0          6m    10.244.1.101   k8s-node1
nginx-deployment-204403116-kmbwr    1/1     Running   0          6m    10.244.1.96    k8s-node1
nginx-deployment-204403116-kzpnf    1/1     Running   0          6m    10.244.1.100   k8s-node1
nginx-deployment-204403116-vbz47    1/1     Running   0          6m    10.244.1.98    k8s-node1
nginx-deployment-204403116-vvh54    1/1     Running   0          6m    10.244.1.99    k8s-node1
ubuntu@k8s-master:~$
```

图 5-28

除非在 nginx.yml 中删除 nodeSelector 设置，然后通过 kubectl apply 重新部署，如图 5-29 所示。

```
ubuntu@k8s-master:~$
ubuntu@k8s-master:~$ kubectl apply -f nginx.yml
deployment "nginx-deployment" configured
ubuntu@k8s-master:~$ kubectl get pod -o wide
NAME                                READY   STATUS        RESTARTS   AGE   IP             NODE
nginx-deployment-204403116-1kk23    0/1     Terminating   0          7m    <none>         k8s-node1
nginx-deployment-204403116-kmbwr    1/1     Terminating   0          7m    10.244.1.96    k8s-node1
nginx-deployment-204403116-kzpnf    0/1     Terminating   0          7m    <none>         k8s-node1
nginx-deployment-204403116-vbz47    0/1     Terminating   0          7m    <none>         k8s-node1
nginx-deployment-204403116-vvh54    0/1     Terminating   0          7m    <none>         k8s-node1
nginx-deployment-2721169382-56mhh   1/1     Running       0          4s    10.244.2.101   k8s-node2
nginx-deployment-2721169382-bf0rd   1/1     Running       0          6s    10.244.2.99    k8s-node2
nginx-deployment-2721169382-cnh8b   1/1     Running       0          4s    10.244.2.100   k8s-node2
nginx-deployment-2721169382-lqhn6   1/1     Running       0          6s    10.244.2.98    k8s-node2
nginx-deployment-2721169382-q61jr   1/1     Running       0          4s    10.244.1.102   k8s-node1
nginx-deployment-2721169382-r983l   1/1     Running       0          6s    10.244.2.97    k8s-node2
ubuntu@k8s-master:~$
```

图 5-29

Kubernetes 会删除之前的 Pod 并调度和运行新的 Pod。

5.2 DaemonSet

Deployment 部署的副本 Pod 会分布在各个 Node 上，每个 Node 都可能运行好几个副本。DaemonSet 的不同之处在于：每个 Node 上最多只能运行一个副本。

DaemonSet 的典型应用场景有：

（1）在集群的每个节点上运行存储 Daemon，比如 glusterd 或 ceph。

（2）在每个节点上运行日志收集 Daemon，比如 flunentd 或 logstash。

（3）在每个节点上运行监控 Daemon，比如 Prometheus Node Exporter 或 collectd。

其实 Kubernetes 自己就在用 DaemonSet 运行系统组件。执行如下命令，如图 5-30 所示。

```
kubectl get daemonset --namespace=kube-system
```

```
ubuntu@k8s-master:~$
ubuntu@k8s-master:~$ kubectl get daemonset --namespace=kube-system
NAME              DESIRED   CURRENT   READY     UP-TO-DATE   AVAILABLE   NODE-SELECTOR
kube-flannel-ds   3         3         3         3            3           beta.kubernetes.io/arch=am
kube-proxy        3         3         3         3            3           <none>
ubuntu@k8s-master:~$
```

图 5-30

DaemonSet kube-flannel-ds 和 kube-proxy 分别负责在每个节点上运行 flannel 和 kube-proxy 组件，如图 5-31 所示。

```
ubuntu@k8s-master:~$
ubuntu@k8s-master:~$ kubectl get pod --namespace=kube-system -o wide
NAME                                  READY   STATUS    RESTARTS   AGE   IP               NODE
etcd-k8s-master                       1/1     Running   1          8d    192.168.56.105   k8s-master
kube-apiserver-k8s-master             1/1     Running   1          8d    192.168.56.105   k8s-master
kube-controller-manager-k8s-master    1/1     Running   1          8d    192.168.56.105   k8s-master
kube-dns-2425271678-51xc4             3/3     Running   6          3d    10.244.1.89      k8s-node1
kube-flannel-ds-cqbpb                 2/2     Running   11         8d    192.168.56.106   k8s-node1
kube-flannel-ds-v0p3x                 2/2     Running   3          8d    192.168.56.105   k8s-master
kube-flannel-ds-xk49w                 2/2     Running   12         8d    192.168.56.107   k8s-node2
kube-proxy-16mg9                      1/1     Running   3          8d    192.168.56.106   k8s-node1
kube-proxy-wc4j0                      1/1     Running   1          8d    192.168.56.105   k8s-master
kube-proxy-xl5gd                      1/1     Running   4          8d    192.168.56.107   k8s-node2
kube-scheduler-k8s-master             1/1     Running   1          8d    192.168.56.105   k8s-master
ubuntu@k8s-master:~$
```

图 5-31

因为 flannel 和 kube-proxy 属于系统组件，需要在命令行中通过--namespace=kube-system 指定 namespace kube-system。若不指定，则只返回默认 namespace default 中的资源。

5.2.1 kube-flannel-ds

下面我们通过分析 kube-flannel-ds 来学习 DaemonSet。

还记得之前是如何部署 flannel 网络的吗？我们执行了如下命令：

```
kubectl apply -f https://raw.githubusercontent.com/coreos/flannel/master/
Documentation/kube-flannel.yml
```

flannel 的 DaemonSet 就定义在 kube-flannel.yml 中，如图 5-32 所示。

```
apiVersion: extensions/v1beta1
kind: DaemonSet  ①
metadata:
  name: kube-flannel-ds
  namespace: kube-system
  labels:
    tier: node
    app: flannel
spec:
  template:
    metadata:
      labels:
        tier: node
        app: flannel
    spec:
      hostNetwork: true  ②
      nodeSelector:
        beta.kubernetes.io/arch: amd64
      containers:  ③
      - name: kube-flannel
        image: quay.io/coreos/flannel:v0.8.0-amd64
        command: [ "/opt/bin/flanneld", "--ip-masq", "--kube-subnet-mgr" ]
      - name: install-cni
        image: quay.io/coreos/flannel:v0.8.0-amd64
        command: [ "/bin/sh", "-c", "set -e -x; cp -f /etc/kube-flannel/cni-conf.json
```

图 5-32

注意：配置文件的完整内容要更复杂一些，为了更好地学习 DaemonSet，这里只保留了最重要的内容。

① DaemonSet 配置文件的语法和结构与 Deployment 几乎完全一样，只是将 kind 设为 DaemonSet。

② hostName 指定 Pod 直接使用的是 Node 网络，相当于 docker run --network=host。考虑到 flannel 需要为集群提供网络连接，这个要求是合理的。

③ containers 定义了运行 flannel 服务的两个容器。

下面我们再来分析另一个 DaemonSet：kube-proxy。

5.2.2 kube-proxy

由于无法拿到 kube-proxy 的 YAML 文件，只能运行如下命令查看配置：

```
kubectl edit daemonset kube-proxy --namespace=kube-system
```

结果如图 5-33 所示。

```
apiVersion: extensions/v1beta1
kind: DaemonSet  ①
metadata:
  labels:
    k8s-app: kube-proxy
  name: kube-proxy
  namespace: kube-system
spec:
  selector:
    matchLabels:
      k8s-app: kube-proxy
  template:
    metadata:
      labels:
        k8s-app: kube-proxy
    spec:
      containers:  ②
      - command:
        - /usr/local/bin/kube-proxy
        - --kubeconfig=/var/lib/kube-proxy/kubeconfig.conf
        - --cluster-cidr=10.244.0.0/16
        image: gcr.io/google_containers/kube-proxy-amd64:v1.7.4
        name: kube-proxy
status:  ③
  currentNumberScheduled: 3
  desiredNumberScheduled: 3
  numberAvailable: 3
  numberMisscheduled: 0
  numberReady: 3
  observedGeneration: 1
  updatedNumberScheduled: 3
```

图 5-33

同样为了便于理解，这里只保留了最重要的信息。

① kind: DaemonSet 指定这是一个 DaemonSet 类型的资源。

② containers 定义了 kube-proxy 的容器。

③ status 是当前 DaemonSet 的运行时状态，这个部分是 kubectl edit 特有的。其实 Kubernetes 集群中每个当前运行的资源都可以通过 kubectl edit 查看其配置和运行状态，比如 kubectl edit deployment nginx-deployment。

5.2.3 运行自己的 DaemonSet

本小节以 Prometheus Node Exporter 为例演示用户如何运行自己的 DaemonSet。

Prometheus 是流行的系统监控方案，Node Exporter 是 Prometheus 的 agent，以 Daemon 的形式运行在每个被监控节点上。

如果是直接在 Docker 中运行 Node Exporter 容器，命令为：

```
docker run -d \
  -v "/proc:/host/proc" \
  -v "/sys:/host/sys" \
```

```
-v "/:/rootfs" \
--net=host \
prom/node-exporter \
--path.procfs /host/proc \
--path.sysfs /host/sys \
--collector.filesystem.ignored-mount-points "^/(sys|proc|dev|host|etc)($|/)"
```

将其转换为 DaemonSet 的 YAML 配置文件 node_exporter.yml，如图 5-34 所示。

```
apiVersion: extensions/v1beta1
kind: DaemonSet
metadata:
  name: node-exporter-daemonset
spec:
  template:
    metadata:
      labels:
        app: prometheus
    spec:
      hostNetwork: true          ①
      containers:
      - name: node-exporter
        image: prom/node-exporter
        imagePullPolicy: IfNotPresent
        command:                 ②
        - /bin/node_exporter
        - --path.procfs
        - /host/proc
        - --path.sysfs
        - /host/sys
        - --collector.filesystem.ignored-mount-points
        - ^/(sys|proc|dev|host|etc)($|/)
        volumeMounts:            ③
        - name: proc
          mountPath: /host/proc
        - name: sys
          mountPath: /host/sys
        - name: root
          mountPath: /rootfs
      volumes:
      - name: proc
        hostPath:
          path: /proc
      - name: sys
        hostPath:
          path: /sys
      - name: root
        hostPath:
          path: /
```

图 5-34

① 直接使用 Host 的网络。

② 设置容器启动命令。

③ 通过 Volume 将 Host 路径 /proc、/sys 和 / 映射到容器中。我们将在后面详细讨论 Volume。

执行 kubectl apply -f node_exporter.yml，如图 5-35 所示。

```
ubuntu@k8s-master:~$
ubuntu@k8s-master:~$ kubectl apply -f node_exporter.yml
daemonset "node-exporter-daemonset" created
ubuntu@k8s-master:~$
ubuntu@k8s-master:~$ kubectl get pod -o wide
NAME                           READY   STATUS    RESTARTS   AGE   IP               NODE
node-exporter-daemonset-b2w0x  1/1     Running   0          6s    192.168.56.107   k8s-node2
node-exporter-daemonset-kvmkr  1/1     Running   0          6s    192.168.56.106   k8s-node1
ubuntu@k8s-master:~$
```

图 5-35

DaemonSet node-exporter-daemonset 部署成功，k8s-node1 和 k8s-node2 上分别运行了一个 node exporter Pod。

5.3 Job

容器按照持续运行的时间可分为两类：服务类容器和工作类容器。

服务类容器通常持续提供服务，需要一直运行，比如 HTTP Server、Daemon 等。工作类容器则是一次性任务，比如批处理程序，完成后容器就退出。

Kubernetes 的 Deployment、ReplicaSet 和 DaemonSet 都用于管理服务类容器；对于工作类容器，我们使用 Job。

先看一个简单的 Job 配置文件 myjob.yml，如图 5-36 所示。

```
apiVersion: batch/v1  ①
kind: Job  ②
metadata:
  name: myjob
spec:
  template:
    metadata:
      name: myjob
    spec:
      containers:
      - name: hello
        image: busybox
        command: ["echo", "hello k8s job! "]
      restartPolicy: Never  ③
```

图 5-36

① batch/v1 是当前 Job 的 apiVersion。
② 指明当前资源的类型为 Job。
③ restartPolicy 指定什么情况下需要重启容器。对于 Job，只能设置为 Never 或者

OnFailure。对于其他 controller（比如 Deployment），可以设置为 Always。

通过 kubectl apply -f myjob.yml 启动 Job，如图 5-37 所示。

```
ubuntu@k8s-master:~$
ubuntu@k8s-master:~$ kubectl apply -f myjob.yml
job "myjob" created
ubuntu@k8s-master:~$
```

图 5-37

通过 kubectl get job 查看 Job 的状态，如图 5-38 所示。

```
ubuntu@k8s-master:~$
ubuntu@k8s-master:~$ kubectl get job
NAME      DESIRED   SUCCESSFUL   AGE
myjob     1         1            12s
ubuntu@k8s-master:~$
```

图 5-38

DESIRED 和 SUCCESSFUL 都为 1，表示按照预期启动了一个 Pod，并且已经成功执行。通过 kubectl get pod 查看 Pod 的状态，如图 5-39 所示。

```
ubuntu@k8s-master:~$
ubuntu@k8s-master:~$ kubectl get pod
No resources found, use --show-all to see completed objects.
ubuntu@k8s-master:~$
ubuntu@k8s-master:~$ kubectl get pod --show-all
NAME           READY   STATUS      RESTARTS   AGE
myjob-nfkxk    0/1     Completed   0          2m
ubuntu@k8s-master:~$
```

图 5-39

因为 Pod 执行完毕后容器已经退出，需要用 --show-all 才能查看 Completed 状态的 Pod。通过 kubectl logs 可以查看 Pod 的标准输出，如图 5-40 所示。

```
ubuntu@k8s-master:~$
ubuntu@k8s-master:~$ kubectl logs myjob-nfkxk
hello k8s job!
ubuntu@k8s-master:~$
```

图 5-40

5.3.1　Pod 失败的情况

以上是 Pod 成功执行的情况，如果 Pod 失败了会怎么样呢？

我们做个试验，修改 myjob.yml，故意引入一个错误，如图 5-41 所示。

```
apiVersion: batch/v1
kind: Job
metadata:
  name: myjob
spec:
  template:
    metadata:
      name: myjob
    spec:
      containers:
      - name: hello
        image: busybox
        command: ["invalid_command", "hello k8s job!"]
      restartPolicy: Never
```

图 5-41

先删除之前的 Job，如图 5-42 所示。

```
ubuntu@k8s-master:~$
ubuntu@k8s-master:~$ kubectl delete -f myjob.yml
job "myjob" deleted
ubuntu@k8s-master:~$
```

图 5-42

运行新的 Job 并查看状态，如图 5-43 所示。

```
ubuntu@k8s-master:~$
ubuntu@k8s-master:~$ kubectl apply -f myjob.yml
job "myjob" created
ubuntu@k8s-master:~$
ubuntu@k8s-master:~$ kubectl get job
NAME    DESIRED   SUCCESSFUL   AGE
myjob   1         0            7s
ubuntu@k8s-master:~$
```

图 5-43

当前 SUCCESSFUL 的 Pod 数量为 0，查看 Pod 的状态，如图 5-44 所示。

```
ubuntu@k8s-master:~$
ubuntu@k8s-master:~$ kubectl get pod --show-all
NAME            READY   STATUS              RESTARTS   AGE
myjob-0x0z0     0/1     ContainerCannotRun  0          11s
myjob-7f693     0/1     ContainerCannotRun  0          9s
myjob-lv7w3     0/1     ContainerCannotRun  0          6s
myjob-nm1lz     0/1     ContainerCannotRun  0          14s
myjob-qgvm0     0/1     ContainerCannotRun  0          4s
myjob-rbs1x     0/1     ContainerCreating   0          1s
ubuntu@k8s-master:~$
```

图 5-44

可以看到有多个 Pod，状态均不正常。通过 kubectl describe pod 查看某个 Pod 的启动日志，如图 5-45 所示。

```
SubObjectPath              Type      Reason                 Message
-------------              ----      ------                 -------
                           Normal    Scheduled              Successfully assigned myjob-0x0z0 to k8s-node2
                           Normal    SuccessfulMountVolume  MountVolume.SetUp succeeded for volume "default-token-k87vh"
spec.containers{hello}     Normal    Pulling                pulling image "busybox"
spec.containers{hello}     Normal    Pulled                 Successfully pulled image "busybox"
spec.containers{hello}     Normal    Created                Created container
spec.containers{hello}     Warning   Failed                 Error: failed to start container "hello": executable not found in $PATH
                           Warning   FailedSync             Error syncing pod
```

图 5-45

日志显示没有可执行程序，符合我们的预期。

下面解释一个现象：为什么 kubectl get pod 会看到这么多个失败的 Pod？

原因是：当第一个 Pod 启动时，容器失败退出，根据 restartPolicy: Never，此失败容器不会被重启，但 Job DESIRED 的 Pod 是 1，目前 SUCCESSFUL 为 0，不满足，所以 Job controller 会启动新的 Pod，直到 SUCCESSFUL 为 1。对于我们这个例子，SUCCESSFUL 永远也到不了 1，所以 Job controller 会一直创建新的 Pod。为了终止这个行为，只能删除 Job，如图 5-46 所示。

图 5-46

如果将 restartPolicy 设置为 OnFailure 会怎么样？下面我们实践一下，修改 myjob.yml 后重新启动，如图 5-47 所示。

图 5-47

Job 的 SUCCESSFUL Pod 数量还是 0，再看看 Pod 的情况，如图 5-48 所示。

图 5-48

这里只有一个 Pod，不过 RESTARTS 为 3，而且不断增加，说明 OnFailure 生效，容器失败后会自动重启。

5.3.2　Job 的并行性

有时我们希望能同时运行多个 Pod，提高 Job 的执行效率。这个可以通过 parallelism 设置，如图 5-49 所示。

```
apiVersion: batch/v1
kind: Job
metadata:
  name: myjob
spec:
  parallelism: 2
  template:
    metadata:
      name: myjob
    spec:
      containers:
      - name: hello
        image: busybox
        command: ["echo", "hello k8s job! "]
      restartPolicy: OnFailure
```

图 5-49

这里我们将并行的 Pod 数量设置为 2，实践一下，如图 5-50 所示。

```
ubuntu@k8s-master:~$
ubuntu@k8s-master:~$ kubectl apply -f myjob.yml
job "myjob" created
ubuntu@k8s-master:~$
ubuntu@k8s-master:~$ kubectl get job
NAME     DESIRED   SUCCESSFUL   AGE
myjob    <none>    2            6s
ubuntu@k8s-master:~$
ubuntu@k8s-master:~$ kubectl get pod --show-all -o wide
NAME           READY   STATUS      RESTARTS   AGE   IP            NODE
myjob-cn4zs    0/1     Completed   0          12s   10.244.2.5    k8s-node2
myjob-vhpzs    0/1     Completed   0          12s   10.244.1.28   k8s-node1
ubuntu@k8s-master:~$
```

图 5-50

Job 一共启动了两个 Pod，而且 AGE 相同，可见是并行运行的。

我们还可以通过 completions 设置 Job 成功完成 Pod 的总数，如图 5-51 所示。

```
apiVersion: batch/v1
kind: Job
metadata:
  name: myjob
spec:
  completions: 6
  parallelism: 2
  template:
    metadata:
      name: myjob
    spec:
      containers:
      - name: hello
        image: busybox
        command: ["echo", "hello k8s job! "]
      restartPolicy: OnFailure
```

图 5-51

上面配置的含义是：每次运行两个 Pod，直到总共有 6 个 Pod 成功完成。实践一下，如图 5-52 所示。

```
ubuntu@k8s-master:~$
ubuntu@k8s-master:~$ kubectl apply -f myjob.yml
job "myjob" created
ubuntu@k8s-master:~$
ubuntu@k8s-master:~$ kubectl get job
NAME      DESIRED   SUCCESSFUL   AGE
myjob     6         6            9s
ubuntu@k8s-master:~$
ubuntu@k8s-master:~$
ubuntu@k8s-master:~$ kubectl get pod --show-all -o wide
NAME           READY   STATUS      RESTARTS   AGE   IP            NODE
myjob-0t4zk    0/1     Completed   0          16s   10.244.2.6    k8s-node2
myjob-79xkx    0/1     Completed   0          11s   10.244.2.8    k8s-node2
myjob-p42lx    0/1     Completed   0          14s   10.244.1.30   k8s-node1
myjob-rfvd7    0/1     Completed   0          13s   10.244.2.7    k8s-node2
myjob-srpg9    0/1     Completed   0          12s   10.244.1.31   k8s-node1
myjob-wl8tt    0/1     Completed   0          16s   10.244.1.29   k8s-node1
ubuntu@k8s-master:~$
```

图 5-52

DESIRED 和 SUCCESSFUL 均为 6，符合预期。如果不指定 completions 和 parallelism，默认值均为 1。

上面的例子只是为了演示 Job 的并行特性，实际用途不大。不过现实中确实存在很多需要并行处理的场景。比如批处理程序，每个副本（Pod）都会从任务池中读取任务并执行，副本越多，执行时间就越短，效率就越高。这种类似的场景都可以用 Job 来实现。

5.3.3 定时 Job

Linux 中有 cron 程序定时执行任务，Kubernetes 的 CronJob 提供了类似的功能，可以定时执行 Job。CronJob 配置文件示例如图 5-53 所示。

```
apiVersion: batch/v2alpha1   ①
kind: CronJob   ②
metadata:
  name: hello
spec:
  schedule: "*/1 * * * *"   ③
  jobTemplate:   ④
    spec:
      template:
        spec:
          containers:
          - name: hello
            image: busybox
            command: ["echo", "hello k8s job!"]
          restartPolicy: OnFailure
```

图 5-53

① batch/v2alpha1 是当前 CronJob 的 apiVersion。

② 指明当前资源的类型为 CronJob。

③ schedule 指定什么时候运行 Job，其格式与 Linux cron 一致。这里 */1 * * * * 的含义是每一分钟启动一次。

④ jobTemplate 定义 Job 的模板，格式与前面的 Job 一致。

接下来通过 kubectl apply 创建 CronJob，如图 5-54 所示。

```
ubuntu@k8s-master:~$
ubuntu@k8s-master:~$ kubectl apply -f cronjob.yml
error: error validating "cronjob.yml": error validating data: the server could not find th
resource; if you choose to ignore these errors, turn validation off with --validate=false
ubuntu@k8s-master:~$
```

图 5-54

失败了。这是因为 Kubernetes 默认没有 enable CronJob 功能，需要在 kube-apiserver 中加入这个功能。方法很简单，修改 kube-apiserver 的配置文件 /etc/kubernetes/manifests/kube-apiserver.yaml，如图 5-55 所示。

```
apiVersion: v1
kind: Pod
metadata:
  annotations:
    scheduler.alpha.kubernetes.io/critical-pod: ""
  creationTimestamp: null
  labels:
    component: kube-apiserver
    tier: control-plane
  name: kube-apiserver
  namespace: kube-system
spec:
  containers:
  - command:
    - kube-apiserver
    - --runtime-config=batch/v2alpha1=true
    - --requestheader-group-headers=X-Remote-Group
    - --requestheader-extra-headers-prefix=X-Remote-Extra-
    - --requestheader-allowed-names=front-proxy-client
    - --kubelet-client-key=/etc/kubernetes/pki/apiserver-kubelet-client
    - --kubelet-preferred-address-types=InternalIP,ExternalIP,Hostname
    - --allow-privileged=true
    - --experimental-bootstrap-token-auth=true
    - --requestheader-username-headers=X-Remote-User
    - --service-account-key-file=/etc/kubernetes/pki/sa.pub
```

图 5-55

kube-apiserver 本身也是一个 Pod，在启动参数中加上 --runtime-config=batch/v2alpha1=true 即可。然后重启 kubelet 服务：

```
systemctl restart kubelet.service
```

kubelet 会重启 kube-apiserver Pod。通过 kubectl api-versions 确认 kube-apiserver 现在已经支持 batch/v2alpha1，如图 5-56 所示。

```
ubuntu@k8s-master:~$
ubuntu@k8s-master:~$ kubectl api-versions
apiextensions.k8s.io/v1beta1
apiregistration.k8s.io/v1beta1
apps/v1beta1
authentication.k8s.io/v1
authentication.k8s.io/v1beta1
authorization.k8s.io/v1
authorization.k8s.io/v1beta1
autoscaling/v1
batch/v1
batch/v2alpha1
certificates.k8s.io/v1beta1
extensions/v1beta1
networking.k8s.io/v1
policy/v1beta1
rbac.authorization.k8s.io/v1alpha1
rbac.authorization.k8s.io/v1beta1
settings.k8s.io/v1alpha1
storage.k8s.io/v1
storage.k8s.io/v1beta1
v1
ubuntu@k8s-master:~$
```

图 5-56

再次创建 CronJob，如图 5-57 所示。

```
ubuntu@k8s-master:~$
ubuntu@k8s-master:~$ kubectl apply -f cronjob.yml
cronjob "hello" created
ubuntu@k8s-master:~$
```

图 5-57

这次成功了。通过 kubectl get cronjob 查看 CronJob 的状态，如图 5-58 所示。

```
ubuntu@k8s-master:~$
ubuntu@k8s-master:~$ kubectl get cronjob
NAME    SCHEDULE      SUSPEND   ACTIVE   LAST-SCHEDULE
hello   */1 * * * *   False     0        Tue, 12 Sep 2017 10:21:00 +0800
ubuntu@k8s-master:~$
```

图 5-58

等待几分钟，然后通过 kubectl get jobs 查看 Job 的执行情况，如图 5-59 所示。

```
ubuntu@k8s-master:~$
ubuntu@k8s-master:~$ kubectl get jobs
NAME                DESIRED   SUCCESSFUL   AGE
hello-1505181600    1         1            5m
hello-1505181660    1         1            4m
hello-1505181720    1         1            3m
hello-1505181780    1         1            2m
hello-1505181840    1         1            1m
hello-1505181900    1         1            3s
ubuntu@k8s-master:~$
```

图 5-59

可以看到每隔一分钟就会启动一个 Job。执行 kubectl logs 可查看某个 Job 的运行日志，如图 5-60 所示。

图 5-60

5.4 小结

运行容器化应用是 Kubernetes 最重要的核心功能。为满足不同的业务需要，Kubernetes 提供了多种 Controller，包括 Deployment、DaemonSet、Job、CronJob 等。本章我们通过实践详细学习了这些 Controller，并讨论了它们的特性和应用场景。

第 6 章
通过 Service 访问 Pod

我们不应该期望 Kubernetes Pod 是健壮的,而是要假设 Pod 中的容器很可能因为各种原因发生故障而死掉。Deployment 等 Controller 会通过动态创建和销毁 Pod 来保证应用整体的健壮性。换句话说,Pod 是脆弱的,但应用是健壮的。

每个 Pod 都有自己的 IP 地址。当 Controller 用新 Pod 替代发生故障的 Pod 时,新 Pod 会分配到新的 IP 地址。这样就产生了一个问题:

如果一组 Pod 对外提供服务(比如 HTTP),它们的 IP 很有可能发生变化,那么客户端如何找到并访问这个服务呢?

Kubernetes 给出的解决方案是 Service。

6.1 创建 Service

Kubernetes Service 从逻辑上代表了一组 Pod,具体是哪些 Pod 则是由 label 来挑选的。Service 有自己的 IP,而且这个 IP 是不变的。客户端只需要访问 Service 的 IP,Kubernetes 则负责建立和维护 Service 与 Pod 的映射关系。无论后端 Pod 如何变化,对客户端不会有任何影响,因为 Service 没有变。

来看个例子,创建下面的这个 Deployment,如图 6-1 所示。

```
apiVersion: apps/v1beta1
kind: Deployment
metadata:
  name: httpd
spec:
  replicas: 3
  template:
    metadata:
      labels:
        run: httpd
    spec:
      containers:
      - name: httpd
        image: httpd
        ports:
        - containerPort: 80
```

图 6-1

我们启动了三个 Pod，运行 httpd 镜像，label 是 run: httpd，Service 将会用这个 label 来挑选 Pod，如图 6-2 所示。

```
ubuntu@k8s-master:~$
ubuntu@k8s-master:~$ kubectl apply -f httpd.yml
deployment "httpd" created
ubuntu@k8s-master:~$
ubuntu@k8s-master:~$ kubectl get pod -o wide
NAME                      READY     STATUS    RESTARTS   AGE    IP            NODE
httpd-741508562-192vp     1/1       Running   0          1m     10.244.4.5    k8s-node1
httpd-741508562-6g4fc     1/1       Running   0          1m     10.244.4.4    k8s-node1
httpd-741508562-6hh9g     1/1       Running   0          1m     10.244.5.4    k8s-node2
ubuntu@k8s-master:~$
```

图 6-2

Pod 分配了各自的 IP，这些 IP 只能被 Kubernetes Cluster 中的容器和节点访问，如图 6-3 所示。

接下来创建 Service，其配置文件如图 6-4 所示。

```
ubuntu@k8s-master:~$
ubuntu@k8s-master:~$ curl 10.244.4.5
<html><body><h1>It works!</h1></body></html>
ubuntu@k8s-master:~$
ubuntu@k8s-master:~$ curl 10.244.5.4
<html><body><h1>It works!</h1></body></html>
ubuntu@k8s-master:~$
```

图 6-3

```
apiVersion: v1        ①
kind: Service         ②
metadata:
  name: httpd-svc     ③
spec:
  selector:
    run: httpd        ④
  ports:
  - protocol: TCP     ⑤
    port: 8080
    targetPort: 80
```

图 6-4

① v1 是 Service 的 apiVersion。
② 指明当前资源的类型为 Service。
③ Service 的名字为 httpd-svc。
④ selector 指明挑选那些 label 为 run: httpd 的 Pod 作为 Service 的后端。
⑤ 将 Service 的 8080 端口映射到 Pod 的 80 端口，使用 TCP 协议。

执行 kubectl apply 创建 Service httpd-svc，如图 6-5 所示。

```
ubuntu@k8s-master:~$
ubuntu@k8s-master:~$ kubectl apply -f httpd-svc.yml
service "httpd-svc" created
ubuntu@k8s-master:~$
ubuntu@k8s-master:~$ kubectl get service
NAME          CLUSTER-IP      EXTERNAL-IP   PORT(S)    AGE
httpd-svc     10.99.229.179   <none>        8080/TCP   7s
kubernetes    10.96.0.1       <none>        443/TCP    2d
ubuntu@k8s-master:~$
```

图 6-5

httpd-svc 分配到一个 CLUSTER-IP 10.99.229.179。可以通过该 IP 访问后端的 httpd Pod，如图 6-6 所示。

```
ubuntu@k8s-master:~$
ubuntu@k8s-master:~$ curl 10.99.229.179:8080
<html><body><h1>It works!</h1></body></html>
ubuntu@k8s-master:~$
```

图 6-6

根据前面的端口映射，这里要使用 8080 端口。另外，除了我们创建的 httpd-svc，还有一个 Service kubernetes，Cluster 内部通过这个 Service 访问 Kubernetes API Server。

通过 kubectl describe 可以查看 httpd-svc 与 Pod 的对应关系，如图 6-7 所示。

```
ubuntu@k8s-master:~$
ubuntu@k8s-master:~$ kubectl describe service httpd-svc
Name:                httpd-svc
Namespace:           default
Labels:              <none>
Annotations:         kubectl.kubernetes.io/last-applied-configuration
:8080,"protocol":"TC...
Selector:            run=httpd
Type:                ClusterIP
IP:                  10.99.229.179
Port:                <unset>  8080/TCP
Endpoints:           10.244.4.4:80,10.244.4.5:80,10.244.5.4:80
Session Affinity:    None
Events:              <none>
ubuntu@k8s-master:~$
```

图 6-7

Endpoints 罗列了三个 Pod 的 IP 和端口。我们知道 Pod 的 IP 是在容器中配置的，那么 Service 的 Cluster IP 又是配置在哪里的呢？CLUSTER-IP 又是如何映射到 Pod IP 的呢？

答案是 iptables。

6.2 Cluster IP 底层实现

Cluster IP 是一个虚拟 IP，是由 Kubernetes 节点上的 iptables 规则管理的。

可以通过 iptables-save 命令打印出当前节点的 iptables 规则，因为输出较多，这里只截取与 httpd-svc Cluster IP 10.99.229.179 相关的信息，如图 6-8 所示。

```
-A KUBE-SERVICES ! -s 10.244.0.0/16 -d 10.99.229.179/32 -p tcp -m comment --comment "default/httpd-svc:
Cluster IP" -m tcp --dport 8080 -j KUBE-MARK-MASQ
-A KUBE-SERVICES -d 10.99.229.179/32 -p tcp -m comment --comment "default/httpd-svc: Cluster IP" -m tcp
--dport 8080 -j KUBE-SVC-RL3JAE4GN7VOGDGP
```

图 6-8

这两条规则的含义是：

（1）如果 Cluster 内的 Pod（源地址来自 10.244.0.0/16）要访问 httpd-svc，则允许。

（2）其他源地址访问 httpd-svc，跳转到规则 KUBE-SVC-RL3JAE4GN7VOG DGP。

KUBE-SVC-RL3JAE4GN7VOGDGP 规则如图 6-9 所示。

```
-A KUBE-SVC-RL3JAE4GN7VOGDGP -m comment -comment "default/httpd-svc: " -m
statistic -mode random -probability 0.33332999982 -j KUBE-SEP-C5KB52P4BBJQ35PH
```

```
-A KUBE-SVC-RL3JAE4GN7VOGDGP -m comment -comment "default/httpd-svc: " -m statistic -mode random -prob
ability 0.33332999982 -j KUBE-SEP-C5KB52P4BBJQ35PH
-A KUBE-SVC-RL3JAE4GN7VOGDGP -m comment -comment "default/httpd-svc: " -m statistic -mode random -prob
ability 0.50000000000 -j KUBE-SEP-HGVKQQZZCF7RV4IT
-A KUBE-SVC-RL3JAE4GN7VOGDGP -m comment -comment "default/httpd-svc: " -j KUBE-SEP-XE25WGVXLHEIRVO5
```

图 6-9

（1）1/3 的概率跳转到规则 KUBE-SEP-C5KB52P4BBJQ35PH。

（2）1/3 的概率（剩下 2/3 的一半）跳转到规则 KUBE-SEP-HGVKQQZZCF7RV4IT。

（3）1/3 的概率跳转到规则 KUBE-SEP-XE25WGVXLHEIRVO5。

上面三个跳转的规则如图 6-10 所示。

```
-A KUBE-SEP-C5KB52P4BBJQ35PH -s 10.244.4.4/32 -m comment --comment "default/httpd-svc:" -j KUBE-MARK-MASQ
-A KUBE-SEP-C5KB52P4BBJQ35PH -p tcp -m comment --comment "default/httpd-svc:" -m tcp -j DNAT --to-destination 10.244.4.4:80
-A KUBE-SEP-HGVKQQZZCF7RV4IT -s 10.244.4.5/32 -m comment --comment "default/httpd-svc:" -j KUBE-MARK-MASQ
-A KUBE-SEP-HGVKQQZZCF7RV4IT -p tcp -m comment --comment "default/httpd-svc:" -m tcp -j DNAT --to-destination 10.244.4.5:80
-A KUBE-SEP-XE25WGVXLHEIRVO5 -s 10.244.5.4/32 -m comment --comment "default/httpd-svc:" -j KUBE-MARK-MASQ
-A KUBE-SEP-XE25WGVXLHEIRVO5 -p tcp -m comment --comment "default/httpd-svc:" -m tcp -j DNAT --to-destination 10.244.5.4:80
```

图 6-10

即将请求分别转发到后端的三个 Pod。通过上面的分析，我们得到结论：iptables 将访问 Service 的流量转发到后端 Pod，而且使用类似轮询的负载均衡策略。

另外，需要补充一点：Cluster 的每一个节点都配置了相同的 iptables 规则，这样就确保了整个 Cluster 都能够通过 Service 的 Cluster IP 访问 Service，如图 6-11 所示。

图 6-11

6.3 DNS 访问 Service

在 Cluster 中，除了可以通过 Cluster IP 访问 Service，Kubernetes 还提供了更为方便的 DNS 访问。

kubeadm 部署时会默认安装 kube-dns 组件，如图 6-12 所示。

```
ubuntu@k8s-master:~$
ubuntu@k8s-master:~$ kubectl get deployment --namespace=kube-system
NAME       DESIRED   CURRENT   UP-TO-DATE   AVAILABLE   AGE
kube-dns   1         1         1            1           5d
ubuntu@k8s-master:~$
```

图 6-12

kube-dns 是一个 DNS 服务器。每当有新的 Service 被创建，kube-dns 会添加该 Service 的 DNS 记录。Cluster 中的 Pod 可以通过 <SERVICE_NAME>.<NAMESPACE_NAME> 访问 Service。

比如可以用 httpd-svc.default 访问 Service httpd-svc，如图 6-13 所示。

```
ubuntu@k8s-master:~$
ubuntu@k8s-master:~$ kubectl run busybox --rm -ti --image=busybox /bin/sh
If you don't see a command prompt, try pressing enter.
/ #
/ # wget httpd-svc.default:8080
Connecting to httpd-svc.default:8080 (10.99.229.179:8080)
index.html           100% |*******************************************|    45  0:00:00 ETA
/ #
```

图 6-13

如上所示，我们在一个临时的 busybox Pod 中验证了 DNS 的有效性。另外，由于这个 Pod 与 httpd-svc 同属于 default namespace，因此可以省略 default 直接用 httpd-svc 访问 Service，如图 6-14 所示。

```
/ #
/ # wget httpd-svc:8080
Connecting to httpd-svc:8080 (10.99.229.179:8080)
index.html           100% |*******************************************|    45  0:00:00 ETA
/ #
```

图 6-14

用 nslookup 查看 httpd-svc 的 DNS 信息，如图 6-15 所示。

```
/ #
/ # nslookup httpd-svc
Server:    10.96.0.10
Address 1: 10.96.0.10 kube-dns.kube-system.svc.cluster.local

Name:      httpd-svc
Address 1: 10.99.229.179 httpd-svc.default.svc.cluster.local
/ #
```

图 6-15

DNS 服务器是 kube-dns.kube-system.svc.cluster.local，这实际上就是 kube-dns 组件，它本身是部署在 kube-system namespace 中的一个 Service。

httpd-svc.default.svc.cluster.local 是 httpd-svc 的完整域名。

如果要访问其他 namespace 中的 Service，就必须带上 namesapce 了。kubectl get namespace 查看已有的 namespace，如图 6-16 所示。

```
ubuntu@k8s-master:~$
ubuntu@k8s-master:~$ kubectl get namespace
NAME          STATUS   AGE
default       Active   5d
kube-public   Active   5d
kube-system   Active   5d
ubuntu@k8s-master:~$
```

图 6-16

在 kube-public 中部署 Service httpd2-svc，配置如图 6-17 所示。

图 6-17

通过 namespace: kube-public 指定资源所属的 namespace。多个资源可以在一个 YAML 文件中定义，用 "---" 分割。执行 kubectl apply 创建资源，如图 6-18 所示。

```
ubuntu@k8s-master:~$
ubuntu@k8s-master:~$ kubectl apply -f httpd2.yml
deployment "httpd2" created
service "httpd2-svc" created
ubuntu@k8s-master:~$
```

图 6-18

查看 kube-public 的 Service，如图 6-19 所示。

```
ubuntu@k8s-master:~$
ubuntu@k8s-master:~$ kubectl get service --namespace=kube-public
NAME          CLUSTER-IP        EXTERNAL-IP    PORT(S)     AGE
httpd2-svc    10.103.198.189    <none>         8080/TCP    2m
ubuntu@k8s-master:~$
```

图 6-19

在 busybox Pod 中访问 httpd2-svc，如图 6-20 所示。

```
ubuntu@k8s-master:~$
ubuntu@k8s-master:~$ kubectl run busybox --rm -ti --image=busybox /bin/sh
If you don't see a command prompt, try pressing enter.
/ #
/ # wget wget httpd2-svc:8080
wget: bad address 'wget'
/ #
/ # wget httpd2-svc.kube-public:8080
Connecting to httpd2-svc.kube-public:8080 (10.103.198.189:8080)
index.html           100% |*******************************************|    45   0:00:00 ETA
/ #
```

图 6-20

因为不属于同一个 namespace，所以必须使用 httpd2-svc.kube-public 才能访问到。

6.4 外网如何访问 Service

除了 Cluster 内部可以访问 Service，很多情况下我们也希望应用的 Service 能够暴露给 Cluster 外部。Kubernetes 提供了多种类型的 Service，默认是 ClusterIP。

（1）ClusterIP

Service 通过 Cluster 内部的 IP 对外提供服务，只有 Cluster 内的节点和 Pod 可访问，这是默认的 Service 类型，前面实验中的 Service 都是 ClusterIP。

（2）NodePort

Service 通过 Cluster 节点的静态端口对外提供服务。Cluster 外部可以通过 <NodeIP>:<NodePort> 访问 Service。

（3）LoadBalancer

Service 利用 cloud provider 特有的 load balancer 对外提供服务，cloud provider 负责将 load balancer 的流量导向 Service。目前支持的 cloud provider 有 GCP、AWS、Azur 等。

下面我们来实践 NodePort，Service httpd-svc 的配置文件修改如图 6-21 所示。

```
apiVersion: v1
kind: Service
metadata:
  name: httpd-svc
spec:
  type: NodePort
  selector:
    run: httpd
  ports:
  - protocol: TCP
    port: 8080
    targetPort: 80
```

图 6-21

添加 type: NodePort，重新创建 httpd-svc，如图 6-22 所示。

```
ubuntu@k8s-master:~$
ubuntu@k8s-master:~$ kubectl apply -f httpd-svc.yml
service "httpd-svc" created
ubuntu@k8s-master:~$
ubuntu@k8s-master:~$ kubectl get service httpd-svc
NAME        CLUSTER-IP       EXTERNAL-IP   PORT(S)          AGE
httpd-svc   10.109.144.35    <nodes>       8080:32312/TCP   5s
ubuntu@k8s-master:~$
```

图 6-22

Kubernetes 依然会为 httpd-svc 分配一个 ClusterIP，不同的是：

（1）EXTERNAL-IP 为 nodes，表示可通过 Cluster 每个节点自身的 IP 访问 Service。

（2）PORT(S) 为 8080:32312。8080 是 ClusterIP 监听的端口，32312 则是节点上监听的端口。Kubernetes 会从 30000 ~ 32767 中分配一个可用的端口，每个节点都会监听此端口并将请求转发给 Service，如图 6-23 所示。

```
ubuntu@k8s-master:~$
ubuntu@k8s-master:~$ netstat -an|grep 32312
tcp6       0      0 :::32312                :::*                    LISTEN
ubuntu@k8s-master:~$
```

图 6-23

下面测试 NodePort 是否正常工作，如图 6-24 所示。

```
ubuntu@k8s-master:~$
ubuntu@k8s-master:~$ curl 192.168.56.105:32312
<html><body><h1>It works!</h1></body></html>
ubuntu@k8s-master:~$
ubuntu@k8s-master:~$ curl 192.168.56.106:32312
<html><body><h1>It works!</h1></body></html>
ubuntu@k8s-master:~$
ubuntu@k8s-master:~$ curl 192.168.56.107:32312
<html><body><h1>It works!</h1></body></html>
ubuntu@k8s-master:~$
```

图 6-24

通过三个节点 IP + 32312 端口都能够访问 httpd-svc。

接下来我们深入探讨一个问题：Kubernetes 是如何将 <NodeIP>:<NodePort> 映射到 Pod

的呢？

与 ClusterIP 一样，也是借助了 iptables。与 ClusterIP 相比，每个节点的 iptables 中都增加了下面两条规则，如图 6-25 所示。

```
-A KUBE-NODEPORTS -p tcp -m comment --comment "default/httpd-svc: " -m tcp --dport 32312 -j KUBE-MARK-MASQ
-A KUBE-NODEPORTS -p tcp -m comment --comment "default/httpd-svc: " -m tcp --dport 32312 -j KUBE-SVC-RL3JAE4GN7VOGDGP
```

图 6-25

规则的含义是：访问当前节点 32312 端口的请求会应用规则 KUBE-SVC-RL3JAE4GN7VOGDGP，内容如图 6-26 所示。

```
-A KUBE-SVC-RL3JAE4GN7VOGDGP -m comment --comment "default/httpd-svc: " -m statistic --mode random --probability 0.33332999982 -j KUBE-SEP-C5KB52P4BBJQ35PH
-A KUBE-SVC-RL3JAE4GN7VOGDGP -m comment --comment "default/httpd-svc: " -m statistic --mode random --probability 0.50000000000 -j KUBE-SEP-HGVKQQZZCF7RV4IT
-A KUBE-SVC-RL3JAE4GN7VOGDGP -m comment --comment "default/httpd-svc: " -j KUBE-SEP-XE25WGVXLHEIRVO5
```

图 6-26

其作用就是负载均衡到每一个 Pod。

NodePort 默认的是随机选择，不过我们可以用 nodePort 指定某个特定端口，如图 6-27 所示。

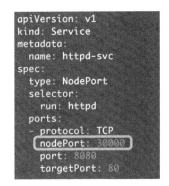

图 6-27

现在配置文件中就有三个 Port 了：

- nodePort 是节点上监听的端口。
- port 是 ClusterIP 上监听的端口。
- targetPort 是 Pod 监听的端口。

最终，Node 和 ClusterIP 在各自端口上接收到的请求都会通过 iptables 转发到 Pod 的 targetPort。

应用新的 nodePort 并验证，如图 6-28 所示。

```
ubuntu@k8s-master:~$
ubuntu@k8s-master:~$ kubectl apply -f httpd-svc.yml
service "httpd-svc" configured
ubuntu@k8s-master:~$
ubuntu@k8s-master:~$ kubectl get service httpd-svc
NAME         CLUSTER-IP       EXTERNAL-IP   PORT(S)            AGE
httpd-svc    10.109.144.35    <nodes>       8080:30000/TCP     36m
ubuntu@k8s-master:~$
ubuntu@k8s-master:~$ curl 192.168.56.105:30000
<html><body><h1>It works!</h1></body></html>
ubuntu@k8s-master:~$ curl 192.168.56.106:30000
<html><body><h1>It works!</h1></body></html>
ubuntu@k8s-master:~$ curl 192.168.56.107:30000
<html><body><h1>It works!</h1></body></html>
ubuntu@k8s-master:~$
```

图 6-28

nodePort: 30000 已经生效了。

6.5 小结

本章我们讨论访问应用的机制 Service，学习了如何创建 Service，Service 的三种类型 ClusterIP、NodePort 和 LoadBalancer，以及它们各自的适用场景。

第 7 章
Rolling Update

滚动更新是一次只更新一小部分副本,成功后再更新更多的副本,最终完成所有副本的更新。滚动更新的最大好处是零停机,整个更新过程始终有副本在运行,从而保证了业务的连续性。

7.1 实践

下面我们部署三副本应用,初始镜像为 httpd:2.2.31,然后将其更新到 httpd:2.2.32。httpd:2.2.31 的配置文件如图 7-1 所示。

```
apiVersion: apps/v1beta1
kind: Deployment
metadata:
  name: httpd
spec:
  replicas: 3
  template:
    metadata:
      labels:
        run: httpd
    spec:
      containers:
      - name: httpd
        image: :httpd:2.2.31
        ports:
        - containerPort: 80
```

图 7-1

通过 kubectl apply 部署,如图 7-2 所示。

```
ubuntu@k8s-master:~$
ubuntu@k8s-master:~$ kubectl apply -f httpd.yml
deployment "httpd" created
ubuntu@k8s-master:~$
ubuntu@k8s-master:~$ kubectl get deployment httpd -o wide
NAME      DESIRED   CURRENT   UP-TO-DATE   AVAILABLE   AGE   CONTAINER(S)   IMAGE(S)       SELECTOR
httpd     3         3         3            3           8s    httpd          httpd:2.2.31   run=httpd
ubuntu@k8s-master:~$ kubectl get replicaset -o wide
NAME             DESIRED   CURRENT   READY   AGE   CONTAINER(S)   IMAGE(S)       SELECTOR
httpd-551879778  3         3         3       13s   httpd          httpd:2.2.31   pod-template-hash=55
ubuntu@k8s-master:~$
ubuntu@k8s-master:~$ kubectl get pod
NAME                     READY   STATUS    RESTARTS   AGE
httpd-551879778-jkxn4    1/1     Running   0          19s
httpd-551879778-n3sqv    1/1     Running   0          19s
httpd-551879778-zdfkt    1/1     Running   0          19s
ubuntu@k8s-master:~$
```

图 7-2

部署过程如下：

（1）创建 Deployment httpd。

（2）创建 ReplicaSet httpd-551879778。

（3）创建三个 Pod。

（4）当前镜像为 httpd:2.2.31。

将配置文件中的 httpd:2.2.31 替换为 httpd:2.2.32，再次执行 kubectl apply，如图 7-3 所示。

```
ubuntu@k8s-master:~$
ubuntu@k8s-master:~$ kubectl apply -f httpd.yml
deployment "httpd" configured
ubuntu@k8s-master:~$
ubuntu@k8s-master:~$ kubectl get deployment httpd -o wide
NAME      DESIRED   CURRENT   UP-TO-DATE   AVAILABLE   AGE   CONTAINER(S)   IMAGE(S)       SELECTOR
httpd     3         3         3            3           1m    httpd          httpd:2.2.32   run=httpd
ubuntu@k8s-master:~$ kubectl get replicaset -o wide
NAME              DESIRED   CURRENT   READY   AGE   CONTAINER(S)   IMAGE(S)       SELECTOR
httpd-1276601241  3         3         3       9s    httpd          httpd:2.2.32   pod-template-hash=
httpd-551879778   0         0         0       2m    httpd          httpd:2.2.31   pod-template-hash=
ubuntu@k8s-master:~$ kubectl get pod
NAME                        READY   STATUS    RESTARTS   AGE
httpd-1276601241-26jx3      1/1     Running   0          13s
httpd-1276601241-27kh7      1/1     Running   0          10s
httpd-1276601241-pwrt7      1/1     Running   0          11s
ubuntu@k8s-master:~$
```

图 7-3

我们发现了如下变化：

（1）Deployment httpd 的镜像更新为 httpd:2.2.32。

（2）新创建了 ReplicaSet httpd-1276601241，镜像为 httpd:2.2.32，并且管理了三个新的 Pod。

（3）之前的 ReplicaSet httpd-551879778 里面已经没有任何 Pod。

结论是：ReplicaSet httpd-551879778 的三个 httpd:2.2.31 Pod 已经被 ReplicaSet httpd-1276601241 的三个 httpd:2.2.32 Pod 替换了。

具体过程可以通过 kubectl describe deployment httpd 查看，如图 7-4 所示。

```
From              SubObjectPath    Type      Reason              Message
----              -------------    ----      ------              -------
deployment-controller              Normal    ScalingReplicaSet   Scaled up replica set httpd-551879778 to 3
deployment-controller              Normal    ScalingReplicaSet   Scaled up replica set httpd-1276601241 to 1
deployment-controller              Normal    ScalingReplicaSet   Scaled down replica set httpd-551879778 to 2
deployment-controller              Normal    ScalingReplicaSet   Scaled up replica set httpd-1276601241 to 2
deployment-controller              Normal    ScalingReplicaSet   Scaled down replica set httpd-551879778 to 1
deployment-controller              Normal    ScalingReplicaSet   Scaled up replica set httpd-1276601241 to 3
deployment-controller              Normal    ScalingReplicaSet   Scaled down replica set httpd-551879778 to 0
```

图 7-4

每次只更新替换一个 Pod：

（1）ReplicaSet httpd-1276601241 增加一个 Pod，总数为 1。

（2）ReplicaSet httpd-551879778 减少一个 Pod，总数为 2。

（3）ReplicaSet httpd-1276601241 增加一个 Pod，总数为 2。

（4）ReplicaSet httpd-551879778 减少一个 Pod，总数为 1。

（5）ReplicaSet httpd-1276601241 增加一个 Pod，总数为 3。

（6）ReplicaSet httpd-551879778 减少一个 Pod，总数为 0。

每次替换的 Pod 数量是可以定制的。Kubernetes 提供了两个参数 maxSurge 和 maxUnavailable 来精细控制 Pod 的替换数量，我们将在后面结合 Health Check 特性一起讨论。

7.2 回滚

kubectl apply 每次更新应用时，Kubernetes 都会记录下当前的配置，保存为一个 revision（版次），这样就可以回滚到某个特定 revision。

默认配置下，Kubernetes 只会保留最近的几个 revision，可以在 Deployment 配置文件中通过 revisionHistoryLimit 属性增加 revision 数量。

下面实践回滚功能。应用有三个配置文件，即 httpd.v1.yml、httpd.v2.yml 和 httpd.v3.yml，分别对应不同的 httpd 镜像 2.4.16、2.4.17 和 2.4.18，如图 7-5、图 7-6、图 7-7 所示。

```
apiVersion: apps/v1beta1
kind: Deployment
metadata:
  name: httpd
spec:
  revisionHistoryLimit: 10
  replicas: 3
  template:
    metadata:
      labels:
        run: httpd
    spec:
      containers:
      - name: httpd
        image: httpd:2.4.16
        ports:
        - containerPort: 80
```

图 7-5

```
apiVersion: apps/v1beta1
kind: Deployment
metadata:
  name: httpd
spec:
  revisionHistoryLimit: 10
  replicas: 3
  template:
    metadata:
      labels:
        run: httpd
    spec:
      containers:
      - name: httpd
        image: httpd:2.4.17
        ports:
        - containerPort: 80
```

图 7-6

```
apiVersion: apps/v1beta1
kind: Deployment
metadata:
  name: httpd
spec:
  revisionHistoryLimit: 10
  replicas: 3
  template:
    metadata:
      labels:
        run: httpd
    spec:
      containers:
      - name: httpd
        image: httpd:2.4.18
        ports:
        - containerPort: 80
```

图 7-7

通过 kubectl apply 部署并更新应用，如图 7-8 所示。

```
ubuntu@k8s-master:~$
ubuntu@k8s-master:~$ kubectl apply -f httpd.v1.yml --record
deployment "httpd" created
ubuntu@k8s-master:~$
ubuntu@k8s-master:~$ kubectl get deployment httpd -o wide
NAME    DESIRED  CURRENT  UP-TO-DATE  AVAILABLE  AGE  CONTAINER(S)  IMAGE(S)      SELECTOR
httpd   3        3        3           3          8s   httpd         httpd:2.4.16  run=httpd
ubuntu@k8s-master:~$
ubuntu@k8s-master:~$ kubectl apply -f httpd.v2.yml --record
deployment "httpd" configured
ubuntu@k8s-master:~$
ubuntu@k8s-master:~$ kubectl get deployment httpd -o wide
NAME    DESIRED  CURRENT  UP-TO-DATE  AVAILABLE  AGE  CONTAINER(S)  IMAGE(S)      SELECTOR
httpd   3        3        3           3          27s  httpd         httpd:2.4.17  run=httpd
ubuntu@k8s-master:~$
ubuntu@k8s-master:~$ kubectl apply -f httpd.v3.yml --record
deployment "httpd" configured
ubuntu@k8s-master:~$
ubuntu@k8s-master:~$ kubectl get deployment httpd -o wide
NAME    DESIRED  CURRENT  UP-TO-DATE  AVAILABLE  AGE  CONTAINER(S)  IMAGE(S)      SELECTOR
httpd   3        3        3           3          51s  httpd         httpd:2.4.18  run=httpd
ubuntu@k8s-master:~$
```

图 7-8

--record 的作用是将当前命令记录到 revision 记录中，这样我们就可以知道每个 revison 对应的是哪个配置文件了。通过 kubectl rollout history deployment httpd 查看 revison 历史记录，如图 7-9 所示。

```
ubuntu@k8s-master:~$
ubuntu@k8s-master:~$ kubectl rollout history deployment httpd
deployments "httpd"
REVISION    CHANGE-CAUSE
1           kubectl apply --filename=httpd.v1.yml --record=true
2           kubectl apply --filename=httpd.v2.yml --record=true
3           kubectl apply --filename=httpd.v3.yml --record=true
ubuntu@k8s-master:~$
```

图 7-9

CHANGE-CAUSE 就是 --record 的结果。如果要回滚到某个版本，比如 revision 1，可以执行命令 kubectl rollout undo deployment httpd --to-revision=1，如图 7-10 所示。

```
ubuntu@k8s-master:~$
ubuntu@k8s-master:~$ kubectl rollout undo deployment httpd --to-revision=1
deployment "httpd" rolled back
ubuntu@k8s-master:~$
ubuntu@k8s-master:~$ kubectl get deployment httpd -o wide
NAME    DESIRED  CURRENT  UP-TO-DATE  AVAILABLE  AGE  CONTAINER(S)  IMAGE(S)      SELECTOR
httpd   3        3        3           3          8m   httpd         httpd:2.4.16  run=httpd
ubuntu@k8s-master:~$
```

图 7-10

此时，revison 历史记录也会发生相应变化，如图 7-11 所示。

```
ubuntu@k8s-master:~$
ubuntu@k8s-master:~$ kubectl rollout history deployment httpd
deployments "httpd"
REVISION    CHANGE-CAUSE
2           kubectl apply --filename=httpd.v2.yml --record=true
3           kubectl apply --filename=httpd.v3.yml --record=true
4           kubectl apply --filename=httpd.v1.yml --record=true
ubuntu@k8s-master:~$
```

图 7-11

revison 1 变成了 revison 4。不过我们可以通过 CHANGE-CAUSE 知道每个 revison 的具体含义，所以一定要在执行 kubectl apply 时加上 --record 参数。

7.3 小结

本章我们学习了滚动更新。滚动更新采用渐进的方式逐步替换旧版本 Pod。如果更新不如预期，可以通过回滚操作恢复到更新前的状态。

第 8 章 Health Check

强大的自愈能力是 Kubernetes 这类容器编排引擎的一个重要特性。自愈的默认实现方式是自动重启发生故障的容器。除此之外，用户还可以利用 Liveness 和 Readiness 探测机制设置更精细的健康检查，进而实现如下需求：

（1）零停机部署。
（2）避免部署无效的镜像。
（3）更加安全的滚动升级。

下面通过实践学习 Kubernetes 的 Health Check 功能。

8.1 默认的健康检查

我们首先学习 Kubernetes 默认的健康检查机制：每个容器启动时都会执行一个进程，此进程由 Dockerfile 的 CMD 或 ENTRYPOINT 指定。如果进程退出时返回码非零，则认为容器发生故障，Kubernetes 就会根据 restartPolicy 重启容器。

下面我们模拟一个容器发生故障的场景，Pod 配置文件如图 8-1 所示。

图 8-1

Pod 的 restartPolicy 设置为 OnFailure，默认为 Always。

sleep 10; exit 1 模拟容器启动 10 秒后发生故障。

执行 kubectl apply 创建 Pod，命名为 healthcheck，如图 8-2 所示。

```
ubuntu@k8s-master:~$
ubuntu@k8s-master:~$ kubectl apply -f healthcheck.yml
pod "healthcheck" created
ubuntu@k8s-master:~$
```

图 8-2

过几分钟查看 Pod 的状态，如图 8-3 所示。

```
ubuntu@k8s-master:~$
ubuntu@k8s-master:~$ kubectl get pod healthcheck
NAME          READY     STATUS      RESTARTS   AGE
healthcheck   1/1       Running     3          1m
ubuntu@k8s-master:~$
```

图 8-3

可看到容器当前已经重启了 3 次。

在上面的例子中，容器进程返回值非零，Kubernetes 则认为容器发生故障，需要重启。有不少情况是发生了故障，但进程并不会退出。比如访问 Web 服务器时显示 500 内部错误，可能是系统超载，也可能是资源死锁，此时 httpd 进程并没有异常退出，在这种情况下重启容器可能是最直接、最有效的解决方案，那我们如何利用 Health Check 机制来处理这类场景呢？

答案就是 Liveness 探测。

8.2 Liveness 探测

Liveness 探测让用户可以自定义判断容器是否健康的条件。如果探测失败，Kubernetes 就会重启容器。

下面举例说明，创建 Pod，如图 8-4 所示。

```
apiVersion: v1
kind: Pod
metadata:
  labels:
    test: liveness
  name: liveness
spec:
  restartPolicy: OnFailure
  containers:
  - name: liveness
    image: busybox
    args:
    - /bin/sh
    - -c
    - touch /tmp/healthy; sleep 30; rm -rf /tmp/healthy; sleep 600
    livenessProbe:
      exec:
        command:
        - cat
        - /tmp/healthy
      initialDelaySeconds: 10
      periodSeconds: 5
```

图 8-4

启动进程首先创建文件 /tmp/healthy，30 秒后删除，在我们的设定中，如果 /tmp/healthy 文件存在，则认为容器处于正常状态，反之则发生故障。

livenessProbe 部分定义如何执行 Liveness 探测：

（1）探测的方法是：通过 cat 命令检查 /tmp/healthy 文件是否存在。如果命令执行成功，返回值为零，Kubernetes 则认为本次 Liveness 探测成功；如果命令返回值非零，本次 Liveness 探测失败。

（2）initialDelaySeconds：10 指定容器启动 10 之后开始执行 Liveness 探测，我们一般会根据应用启动的准备时间来设置。比如某个应用正常启动要花 30 秒，那么 initialDelaySeconds 的值就应该大于 30。

（3）periodSeconds：5 指定每 5 秒执行一次 Liveness 探测。Kubernetes 如果连续执行 3 次 Liveness 探测均失败，则会杀掉并重启容器。

下面创建 Pod liveness，如图 8-5 所示。

```
ubuntu@k8s-master:~$
ubuntu@k8s-master:~$ kubectl apply -f liveness.yml
pod "liveness" created
ubuntu@k8s-master:~$
```

图 8-5

从配置文件可知，最开始的 30 秒，/tmp/healthy 存在，cat 命令返回 0，Liveness 探测成功，这段时间 kubectl describe pod liveness 的 Events 部分会显示正常的日志，如图 8-6 所示。

```
Type        Reason                  Message
--------    ------                  -------
Normal      Scheduled               Successfully assigned liveness to k8s-node1
Normal      SuccessfulMountVolume   MountVolume.SetUp succeeded for volume "default-token-hnz7b"
Normal      Pulling                 pulling image "busybox"
Normal      Pulled                  Successfully pulled image "busybox"
Normal      Created                 Created container
Normal      Started                 Started container
```

图 8-6

35 秒之后，日志会显示 /tmp/healthy 已经不存在，Liveness 探测失败。再过几十秒，几次探测都失败后，容器会被重启，如图 8-7、图 8-8 所示。

```
                Type        Reason                  Message
                --------    ------                  -------
                Normal      Scheduled               Successfully assigned liveness to k8s-node1
                Normal      SuccessfulMountVolume   MountVolume.SetUp succeeded for volume "default-token-hnz7b"
{liveness}      Warning     Unhealthy               Liveness probe failed: cat: can't open '/tmp/healthy': No such file

Normal  Pulling pulling image "busybox"
Normal  Killing Killing container with id docker://liveness:pod "liveness_default(308650ed-a32f-11e7-b2d9-0800274451ad)" container "liveness" is unhealthy, it will be killed and re-created.
Normal  Pulled  Successfully pulled image "busybox"
Normal  Created Created container
Normal  Started Started container
```

图 8-7

```
ubuntu@k8s-master:~$
ubuntu@k8s-master:~$ kubectl get pod liveness
NAME       READY    STATUS    RESTARTS   AGE
liveness   1/1      Running   1          1m
ubuntu@k8s-master:~$
```

图 8-8

8.3 Readiness 探测

除了 Liveness 探测，Kubernetes Health Check 机制还包括 Readiness 探测。

用户通过 Liveness 探测可以告诉 Kubernetes 什么时候通过重启容器实现自愈；Readiness 探测则是告诉 Kubernetes 什么时候可以将容器加入到 Service 负载均衡池中，对外提供服务。

Readiness 探测的配置语法与 Liveness 探测完全一样，如图 8-9 中的例子所示。

```
apiVersion: v1
kind: Pod
metadata:
  labels:
    test: readiness
  name: readiness
spec:
  restartPolicy: OnFailure
  containers:
  - name: readiness
    image: busybox
    args:
    - /bin/sh
    - -c
    - touch /tmp/healthy; sleep 30; rm -rf /tmp/healthy; sleep 600
    readinessProbe:
      exec:
        command:
        - cat
        - /tmp/healthy
      initialDelaySeconds: 10
      periodSeconds: 5
```

图 8-9

这个配置文件只是将前面例子中的 liveness 替换为了 readiness，我们看看有什么不同的效果，如图 8-10 所示。

```
ubuntu@k8s-master:~$ kubectl apply -f readiness.y
pod "readiness" created
ubuntu@k8s-master:~$
ubuntu@k8s-master:~$ kubectl get pod readiness
NAME        READY   STATUS    RESTARTS   AGE
readiness   0/1     Running   0          9s
ubuntu@k8s-master:~$
ubuntu@k8s-master:~$ kubectl get pod readiness
NAME        READY   STATUS    RESTARTS   AGE
readiness   1/1     Running   0          16s
ubuntu@k8s-master:~$
ubuntu@k8s-master:~$ kubectl get pod readiness
NAME        READY   STATUS    RESTARTS   AGE
readiness   0/1     Running   0          53s
ubuntu@k8s-master:~$
```

图 8-10

Pod readiness 的 READY 状态经历了如下变化：

（1）刚被创建时，READY 状态为不可用。

（2）15 秒后（initialDelaySeconds + periodSeconds），第一次进行 Readiness 探测并成功返回，设置 READY 为可用。

（3）30 秒后，/tmp/healthy 被删除，连续 3 次 Readiness 探测均失败后，READY 被设置为不可用。

通过 kubectl describe pod readiness 也可以看到 Readiness 探测失败的日志，如图 8-11 所示。

```
Type      Reason                 Message
----      ------                 -------
Normal    Scheduled              Successfully assigned readiness to k8s-node1
Normal    SuccessfulMountVolume  MountVolume.SetUp succeeded for volume "default-token-hnz7b"
Normal    Pulling                pulling image "busybox"
Normal    Pulled                 Successfully pulled image "busybox"
Normal    Created                Created container
Normal    Started                Started container
Warning   Unhealthy              Readiness probe failed: cat: can't open '/tmp/healthy': No such file
```

图 8-11

下面对 Liveness 探测和 Readiness 探测做个比较：

（1）Liveness 探测和 Readiness 探测是两种 Health Check 机制，如果不特意配置，Kubernetes 将对两种探测采取相同的默认行为，即通过判断容器启动进程的返回值是否为零来判断探测是否成功。

（2）两种探测的配置方法完全一样，支持的配置参数也一样。不同之处在于探测失败后的行为：Liveness 探测是重启容器；Readiness 探测则是将容器设置为不可用，不接收 Service 转发的请求。

（3）Liveness 探测和 Readiness 探测是独立执行的，二者之间没有依赖，所以可以单独使用，也可以同时使用。用 Liveness 探测判断容器是否需要重启以实现自愈；用 Readiness 探测判断容器是否已经准备好对外提供服务。

理解了 Liveness 探测和 Readiness 探测的原理，接下来讨论如何在业务场景中使用 Health Check。

8.4 Health Check 在 Scale Up 中的应用

对于多副本应用，当执行 Scale Up 操作时，新副本会作为 backend 被添加到 Service 的负载均衡中，与已有副本一起处理客户的请求。考虑到应用启动通常都需要一个准备阶段，比如加载缓存数据、连接数据库等，从容器启动到真正能够提供服务是需要一段时间的。我们可以通过 Readiness 探测判断容器是否就绪，避免将请求发送到还没有准备好的 backend。

示例应用的配置文件如图 8-12 所示。

```
apiVersion: apps/v1beta1
kind: Deployment
metadata:
  name: web
spec:
  replicas: 3
  template:
    metadata:
      labels:
        run: web
    spec:
      containers:
      - name: web
        image: myhttpd
        ports:
        - containerPort: 8080
        readinessProbe:
          httpGet:
            scheme: HTTP
            path: /healthy
            port: 8080
          initialDelaySeconds: 10
          periodSeconds: 5

---
apiVersion: v1
kind: Service
metadata:
  name: web-svc
spec:
  selector:
    run: web
  ports:
  - protocol: TCP
    port: 8080
    targetPort: 80
```

图 8-12

重点关注 readinessProbe 部分。这里我们使用了不同于 exec 的另一种探测方法 httpGet。Kubernetes 对于该方法探测成功的判断条件是 http 请求的返回代码在 200～400 之间。

- schema 指定协议，支持 HTTP（默认值）和 HTTPS。

- path 指定访问路径。
- port 指定端口。

上面配置的作用是：

（1）容器启动 10 秒之后开始探测。

（2）如果 http://[container_ip]:8080/healthy 返回代码不是 200~400，表示容器没有就绪，不接收 Service web-svc 的请求。

（3）每隔 5 秒探测一次。

（4）直到返回代码为 200~400，表明容器已经就绪，然后将其加入到 web-svc 的负载均衡中，开始处理客户请求。

（5）探测会继续以 5 秒的间隔执行，如果连续发生 3 次失败，容器又会从负载均衡中移除，直到下次探测成功重新加入。

对于 http://[container_ip]:8080/healthy，应用则可以实现自己的判断逻辑，比如检查所依赖的数据库是否就绪，示例代码如图 8-13 所示。

```
http.HandleFunc("/healthy", func(w http.ResponseWriter, r *http.Request) {  ①
    healthy = True;
    // Check Database
    db = connect(dbIP, dbPort, dbUser, dbPassword)  ②
    if db != NULL {
        try {
            db.Query("SELECT test;")
        } catch (e){
            err = e.message
        }
    }
    if db == NULL || err != NULL {
        healthy = False
        errMsg += "Database is not ready."
    }
    if healthy {
        w.Write([]byte("OK"))  ③
    } else {
        // Send 503
        http.Error(w, errMsg, http.StatusServiceUnavailable)  ④
    }
})
http.ListenAndServe("8080")  ⑤
```

图 8-13

① 定义 /healthy 的处理函数。

② 连接数据库并执行测试 SQL。

③ 测试成功，正常返回，代码 200。

④ 测试失败，返回错误代码 503。

⑤ 在 8080 端口监听。

对于生产环境中重要的应用，都建议配置 Health Check，保证处理客户请求的容器都是准备就绪的 Service backend。

8.5 Health Check 在滚动更新中的应用

Health Check 另一个重要的应用场景是 Rolling Update。试想一下，现有一个正常运行的多副本应用，接下来对应用进行更新（比如使用更高版本的 image），Kubernetes 会启动新副本，然后发生了如下事件：

（1）正常情况下新副本需要 10 秒钟完成准备工作，在此之前无法响应业务请求。
（2）由于人为配置错误，副本始终无法完成准备工作（比如无法连接后端数据库）。

先别继续往下看，现在请花一分钟思考这个问题：如果没有配置 Health Check，会出现怎样的情况？

因为新副本本身没有异常退出，默认的 Health Check 机制会认为容器已经就绪，进而会逐步用新副本替换现有副本，其结果就是：当所有旧副本都被替换后，整个应用将无法处理请求，无法对外提供服务。如果这是发生在重要的生产系统上，后果会非常严重。

如果正确配置了 Health Check，新副本只有通过了 Readiness 探测才会被添加到 Service；如果没有通过探测，现有副本不会被全部替换，业务仍然正常进行。

下面通过例子来实践 Health Check 在 Rolling Update 中的应用。

使用如下配置文件 app.v1.yml 模拟一个 10 副本的应用，如图 8-14 所示。

```
apiVersion: apps/v1beta1
kind: Deployment
metadata:
  name: app
spec:
  replicas: 10
  template:
    metadata:
      labels:
        run: app
    spec:
      containers:
      - name: app
        image: busybox
        args:
        - /bin/sh
        - -c
        - sleep 10; touch /tmp/healthy; sleep 30000
        readinessProbe:
          exec:
            command:
            - cat
            - /tmp/healthy
          initialDelaySeconds: 10
          periodSeconds: 5
```

图 8-14

10 秒后副本能够通过 Readiness 探测，如图 8-15 所示。

```
ubuntu@k8s-master:~$
ubuntu@k8s-master:~$ kubectl apply -f app.v1.yml --record
deployment "app" created
ubuntu@k8s-master:~$
ubuntu@k8s-master:~$ kubectl get deployment app
NAME      DESIRED   CURRENT   UP-TO-DATE   AVAILABLE   AGE
app       10        10        10           10          28s
ubuntu@k8s-master:~$
ubuntu@k8s-master:~$ kubectl get pod
NAME                     READY    STATUS    RESTARTS   AGE
app-2780995820-0mpfl     1/1      Running   0          32s
app-2780995820-9nmfm     1/1      Running   0          32s
app-2780995820-dqdwn     1/1      Running   0          32s
app-2780995820-g0srs     1/1      Running   0          32s
app-2780995820-g52wp     1/1      Running   0          32s
app-2780995820-kddms     1/1      Running   0          32s
app-2780995820-rrwsh     1/1      Running   0          32s
app-2780995820-t3kl4     1/1      Running   0          32s
app-2780995820-v1qzn     1/1      Running   0          32s
app-2780995820-z8qx4     1/1      Running   0          32s
ubuntu@k8s-master:~$
```

图 8-15

接下来滚动更新应用，配置文件 app.v2.yml，如图 8-16 所示。

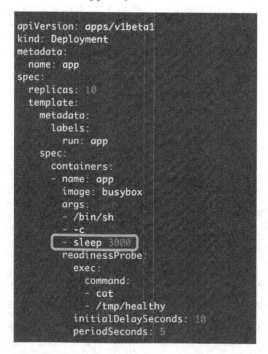

图 8-16

很显然，由于新副本中不存在 /tmp/healthy，因此是无法通过 Readiness 探测的，验证如图 8-17 所示。

```
ubuntu@k8s-master:~$
ubuntu@k8s-master:~$ kubectl apply -f app.v2.yml --record
deployment "app" configured
ubuntu@k8s-master:~$
ubuntu@k8s-master:~$ kubectl get deployment app
NAME      DESIRED   CURRENT   UP-TO-DATE   AVAILABLE   AGE
app       10        13        5            8           5m
ubuntu@k8s-master:~$
ubuntu@k8s-master:~$ kubectl get pod
NAME                     READY   STATUS    RESTARTS   AGE
app-2780995820-0mpfl     1/1     Running   0          5m
app-2780995820-g0srs     1/1     Running   0          5m
app-2780995820-g52wp     1/1     Running   0          5m
app-2780995820-kddms     1/1     Running   0          5m
app-2780995820-rrwsh     1/1     Running   0          5m
app-2780995820-t3kl4     1/1     Running   0          5m
app-2780995820-v1qzn     1/1     Running   0          5m
app-2780995820-z8qx4     1/1     Running   0          5m
app-3350497563-d3ls3     0/1     Running   0          49s
app-3350497563-fkjvq     0/1     Running   0          49s
app-3350497563-ltjp3     0/1     Running   0          49s
app-3350497563-qm92c     0/1     Running   0          49s
app-3350497563-vh56z     0/1     Running   0          49s
ubuntu@k8s-master:~$
```

图 8-17

这个截图包含了大量的信息，值得我们详细分析。

先关注 kubectl get pod 输出：

（1）从 Pod 的 AGE 栏可判断，最后 5 个 Pod 是新副本，目前处于 NOT READY 状态。
（2）旧副本从最初 10 个减少到 8 个。

再来看 kubectl get deployment app 的输出：

（1）DESIRED 10 表示期望的状态是 10 个 READY 的副本。
（2）CURRENT 13 表示当前副本的总数，即 8 个旧副本+5 个新副本。
（3）UP-TO-DATE 5 表示当前已经完成更新的副本数，即 5 个新副本。
（4）AVAILABLE 8 表示当前处于 READY 状态的副本数，即 8 个旧副本。

在我们的设定中，新副本始终都无法通过 Readiness 探测，所以这个状态会一直保持下去。

上面我们模拟了一个滚动更新失败的场景。不过幸运的是：Health Check 帮我们屏蔽了有缺陷的副本，同时保留了大部分旧副本，业务没有因更新失败受到影响。

接下来我们要回答：为什么新创建的副本数是 5 个，同时只销毁了 2 个旧副本？

原因是：滚动更新通过参数 maxSurge 和 maxUnavailable 来控制副本替换的数量。

1. maxSurge

此参数控制滚动更新过程中副本总数超过 DESIRED 的上限。maxSurge 可以是具体的整数（比如 3），也可以是百分百，向上取整。maxSurge 默认值为 25%。

在上面的例子中，DESIRED 为 10，那么副本总数的最大值为 roundUp(10 + 10 * 25%) = 13，所以我们看到 CURRENT 就是 13。

2. maxUnavailable

此参数控制滚动更新过程中，不可用的副本相占 DESIRED 的最大比例。maxUnavailable

可以是具体的整数（比如 3），也可以是百分百，向下取整。maxUnavailable 默认值为 25%。

在上面的例子中，DESIRED 为 10，那么可用的副本数至少要为 10 - roundDown(10 * 25%) = 8，所以我们看到 AVAILABLE 是 8。

maxSurge 值越大，初始创建的新副本数量就越多；maxUnavailable 值越大，初始销毁的旧副本数量就越多。

理想情况下，我们这个案例滚动更新的过程应该是这样的：

（1）创建 3 个新副本使副本总数达到 13 个。
（2）销毁 2 个旧副本使可用的副本数降到 8 个。
（3）当 2 个旧副本成功销毁后，再创建 2 个新副本，使副本总数保持为 13 个。
（4）当新副本通过 Readiness 探测后，会使可用副本数增加，超过 8。
（5）进而可以继续销毁更多的旧副本，使可用副本数回到 8。
（6）旧副本的销毁使副本总数低于 13，这样就允许创建更多的新副本。
（7）这个过程会持续进行，最终所有的旧副本都会被新副本替换，滚动更新完成。

而我们的实际情况是在第 4 步就卡住了，新副本无法通过 Readiness 探测。这个过程可以在 kubectl describe deployment app 的日志部分查看，如图 8-18 所示。

图 8-18

如果滚动更新失败，可以通过 kubectl rollout undo 回滚到上一个版本，如图 8-19 所示。

图 8-19

如果要定制 maxSurge 和 maxUnavailable，可以进行如图 8-20 所示的配置。

```
apiVersion: apps/v1beta1
kind: Deployment
metadata:
  name: app
spec:
  strategy:
    rollingUpdate:
      maxSurge: 35%
      maxUnavailable: 35%
  replicas: 10
  template:
    metadata:
      labels:
        run: app
    spec:
      containers:
      - name: app
        image: busybox
        args:
        - /bin/sh
        - -c
        - sleep 3000
        readinessProbe:
          exec:
            command:
            - cat
            - /tmp/healthy
          initialDelaySeconds: 10
          periodSeconds: 5
```

图 8-20

8.6 小结

本章我们讨论了 Kubernetes 健康检查的两种机制：Liveness 探测和 Readiness 探测，并实践了健康检查在 Scale Up 和 Rolling Update 场景中的应用。

第 9 章 数据管理

本章将讨论 Kubernetes 如何管理存储资源。

首先我们会学习 Volume，以及 Kubernetes 如何通过 Volume 为集群中的容器提供存储；然后我们会实践几种常用的 Volume 类型并理解它们各自的应用场景；最后，我们会讨论 Kubernetes 如何通过 Persistent Volume 和 Persistent Volume Claim 分离集群管理员与集群用户的职责，并实践 Volume 的静态供给和动态供给。

9.1 Volume

本节我们讨论 Kubernetes 的存储模型 Volume，学习如何将各种持久化存储映射到容器。

我们经常会说：容器和 Pod 是短暂的。其含义是它们的生命周期可能很短，会被频繁地销毁和创建。容器销毁时，保存在容器内部文件系统中的数据都会被清除。

为了持久化保存容器的数据，可以使用 Kubernetes Volume。

Volume 的生命周期独立于容器，Pod 中的容器可能被销毁和重建，但 Volume 会被保留。

本质上，Kubernetes Volume 是一个目录，这一点与 Docker Volume 类似。当 Volume 被 mount 到 Pod，Pod 中的所有容器都可以访问这个 Volume。Kubernetes Volume 也支持多种 backend 类型，包括 emptyDir、hostPath、GCE Persistent Disk、AWS Elastic Block Store、NFS、Ceph 等，完整列表可参考 https://kubernetes.io/docs/concepts/storage/volumes/#types-of-volumes。

Volume 提供了对各种 backend 的抽象，容器在使用 Volume 读写数据的时候不需要关心数据到底是存放在本地节点的文件系统中还是云硬盘上。对它来说，所有类型的 Volume 都只是一个目录。

我们将从最简单的 emptyDir 开始学习 Kubernetes Volume。

9.1.1 emptyDir

emptyDir 是最基础的 Volume 类型。正如其名字所示，一个 emptyDir Volume 是 Host 上的一个空目录。

emptyDir Volume 对于容器来说是持久的，对于 Pod 则不是。当 Pod 从节点删除时，Volume 的内容也会被删除。但如果只是容器被销毁而 Pod 还在，则 Volume 不受影响。

也就是说：emptyDir Volume 的生命周期与 Pod 一致。

Pod 中的所有容器都可以共享 Volume，它们可以指定各自的 mount 路径。下面通过例子来实践 emptyDir，配置文件如图 9-1 所示。

```
apiVersion: v1
kind: Pod
metadata:
  name: producer-consumer
spec:
  containers:
  - image: busybox
    name: producer
    volumeMounts:
    - mountPath: /producer_dir   ②
      name: shared-volume
    args:
    - /bin/sh   ③
    - -c
    - echo "hello world" > /producer_dir/hello ; sleep 30000

  - image: busybox
    name: consumer
    volumeMounts:
    - mountPath: /consumer_dir   ④
      name: shared-volume
    args:
    - /bin/sh   ⑤
    - -c
    - cat /consumer_dir/hello ; sleep 30000

  volumes:
  - name: shared-volume   ①
    emptyDir: {}
```

图 9-1

这里我们模拟了一个 producer-consumer 场景。Pod 有两个容器 producer 和 consumer，它们共享一个 Volume。producer 负责往 Volume 中写数据，consumer 则是从 Volume 读取数据。

① 文件最底部 volumes 定义了一个 emptyDir 类型的 Volume shared-volume。
② producer 容器将 shared-volume mount 到 /producer_dir 目录。
③ producer 通过 echo 将数据写到文件 hello 里。
④ consumer 容器将 shared-volume mount 到 /consumer_dir 目录。
⑤ consumer 通过 cat 从文件 hello 读数据。

执行命令创建 Pod，如图 9-2 所示。

```
ubuntu@k8s-master:~$
ubuntu@k8s-master:~$ kubectl apply -f emptyDir.yml
pod "producer-consumer" created
ubuntu@k8s-master:~$
ubuntu@k8s-master:~$ kubectl get pod
NAME                READY   STATUS    RESTARTS   AGE
producer-consumer   2/2     Running   0          15s
ubuntu@k8s-master:~$
ubuntu@k8s-master:~$ kubectl logs producer-consumer consumer
hello world
```

图 9-2

kubectl logs 显示容器 consumer 成功读到了 producer 写入的数据，验证了两个容器共享 emptyDir Volume。

因为 emptyDir 是 Docker Host 文件系统里的目录，其效果相当于执行了 docker run -v /producer_dir 和 docker run -v /consumer_dir。通过 docker inspect 查看容器的详细配置信息，我们发现两个容器都 mount 了同一个目录，如图 9-3、图 9-4 所示。

```
"mounts": [
    {
        "Source": "/var/lib/kubelet/pods/3e6100eb-a97a-11e7-8f72-080027
4451ad/volumes/kubernetes.io~empty-dir/shared-volume",
        "Destination": "/producer_dir",
        "Mode": "",
        "RW": true,
        "Propagation": "rprivate"
    },
```

图 9-3

```
"mounts": [
    {
        "Source": "/var/lib/kubelet/pods/3e6100eb-a97a-11e7-8f72-080027
4451ad/volumes/kubernetes.io~empty-dir/shared-volume",
        "Destination": "/consumer_dir",
        "Mode": "",
        "RW": true,
        "Propagation": "rprivate"
    },
```

图 9-4

这里/var/lib/kubelet/pods/3e6100eb-a97a-11e7-8f72-0800274451ad/volumes/kubernetes.io~empty-dir/shared-volume 就是 emptyDir 在 Host 上的真正路径。

emptyDir 是 Host 上创建的临时目录，其优点是能够方便地为 Pod 中的容器提供共享存储，不需要额外的配置。它不具备持久性，如果 Pod 不存在了，emptyDir 也就没有了。根据这个特性，emptyDir 特别适合 Pod 中的容器需要临时共享存储空间的场景，比如前面的生产者消费者用例。

9.1.2 hostPath

hostPath Volume 的作用是将 Docker Host 文件系统中已经存在的目录 mount 给 Pod 的容器。大部分应用都不会使用 hostPath Volume，因为这实际上增加了 Pod 与节点的耦合，限制了 Pod 的使用。不过那些需要访问 Kubernetes 或 Docker 内部数据（配置文件和二进制库）的应用则需要使用 hostPath。

比如 kube-apiserver 和 kube-controller-manager 就是这样的应用，通过 kubectl edit --namespace=kube-system pod kube-apiserver-k8s-master 查看 kube-apiserver Pod 的配置，Volume 的相关部分如图 9-5 所示。

```
      volumeMounts:
      - mountPath: /etc/kubernetes
        name: k8s
        readOnly: true
      - mountPath: /etc/ssl/certs
        name: certs
      - mountPath: /etc/pki
        name: pki
      dnsPolicy: ClusterFirst
      hostNetwork: true
      nodeName: k8s-master
      restartPolicy: Always
      schedulerName: default-scheduler
      securityContext: {}
      terminationGracePeriodSeconds: 30
      tolerations:
      - effect: NoExecute
        operator: Exists
      volumes:
      - hostPath:
          path: /etc/kubernetes
        name: k8s
      - hostPath:
          path: /etc/ssl/certs
        name: certs
      - hostPath:
          path: /etc/pki
        name: pki
```

图 9-5

这里定义了三个 hostPath：volume k8s、certs 和 pki，分别对应 Host 目录/etc/kubernetes、/etc/ssl/certs 和/etc/pki。

如果 Pod 被销毁了，hostPath 对应的目录还是会被保留，从这一点来看，hostPath 的持久性比 emptyDir 强。不过一旦 Host 崩溃，hostPath 也就无法访问了。

接下来我们将学习具备真正持久性的 Volume。

9.1.3　外部 Storage Provider

如果 Kubernetes 部署在诸如 AWS、GCE、Azure 等公有云上，可以直接使用云硬盘作为 Volume。下面给出一个 AWS Elastic Block Store 的例子，如图 9-6 所示。

```yaml
apiVersion: v1
kind: Pod
metadata:
  name: using-ebs
spec:
  containers:
  - image: busybox
    name: using-ebs
    volumeMounts:
    - mountPath: /test-ebs
      name: ebs-volume
  volumes:
  - name: ebs-volume
    # This AWS EBS volume must already exist.
    awsElasticBlockStore:
      volumeID: <volume-id>
      fsType: ext4
```

图 9-6

要在 Pod 中使用 ESB volume，必须先在 AWS 中创建，然后通过 volume-id 引用。其他云硬盘的使用方法可参考各公有云厂商的官方文档。

Kubernetes Volume 也可以使用主流的分布式存储，比如 Ceph、GlusterFS 等。下面给出一个 Ceph 的例子，如图 9-7 所示。

```yaml
apiVersion: v1
kind: Pod
metadata:
  name: using-ceph
spec:
  containers:
    - image: busybox
      name: using-ceph
      volumeMounts:
      - name: ceph-volume
        mountPath: /test-ceph
  volumes:
    - name: ceph-volume
      cephfs:
        path: /some/path/in/side/cephfs
        monitors: "10.16.154.78:6789"
        secretFile: "/etc/ceph/admin.secret"
```

图 9-7

Ceph 文件系统的 /some/path/in/side/cephfs 目录被 mount 到容器路径 /test-ceph。

相对于 emptyDir 和 hostPath，这些 Volume 类型的最大特点就是不依赖 Kubernetes。Volume 的底层基础设施由独立的存储系统管理，与 Kubernetes 集群是分离的。数据被持久化后，即使整个 Kubernetes 崩溃也不会受损。

当然，运维这样的存储系统通常不是一项简单的工作，特别是对可靠性、可用性和扩展性有较高要求的时候。

9.2 PersistentVolume & PersistentVolumeClaim

Volume 提供了非常好的数据持久化方案，不过在可管理性上还有不足。

拿前面的 AWS EBS 例子来说，要使用 Volume，Pod 必须事先知道如下信息：

（1）当前 Volume 来自 AWS EBS。

（2）EBS Volume 已经提前创建，并且知道确切的 volume-id。

Pod 通常是由应用的开发人员维护，而 Volume 则通常是由存储系统的管理员维护。开发人员要获得上面的信息，要么询问管理员，要么自己就是管理员。

这样就带来一个管理上的问题：应用开发人员和系统管理员的职责耦合在一起了。如果系统规模较小或者对于开发环境，这样的情况还可以接受，当集群规模变大，特别是对于生成环境，考虑到效率和安全性，这就成了必须要解决的问题。

Kubernetes 给出的解决方案是 PersistentVolume 和 PersistentVolumeClaim。

PersistentVolume（PV）是外部存储系统中的一块存储空间，由管理员创建和维护。与 Volume 一样，PV 具有持久性，生命周期独立于 Pod。

PersistentVolumeClaim（PVC）是对 PV 的申请（Claim）。PVC 通常由普通用户创建和维护。需要为 Pod 分配存储资源时，用户可以创建一个 PVC，指明存储资源的容量大小和访问模式（比如只读）等信息，Kubernetes 会查找并提供满足条件的 PV。

有了 PersistentVolumeClaim，用户只需要告诉 Kubernetes 需要什么样的存储资源，而不必关心真正的空间从哪里分配、如何访问等底层细节信息。这些 Storage Provider 的底层信息交给管理员来处理，只有管理员才应该关心创建 PersistentVolume 的细节信息。

Kubernetes 支持多种类型的 PersistentVolume，比如 AWS EBS、Ceph、NFS 等，完整列表请参考 https://kubernetes.io/docs/concepts/storage/persistent-volumes/#types-of-persistent-volumes。

下面我们用 NFS 来体会 PersistentVolume 的使用方法。

9.2.1 NFS PersistentVolume

作为准备工作，我们已经在 k8s-master 节点上搭建了一个 NFS 服务器，目录为 /nfsdata，如图 9-8 所示。

```
root@k8s-master:~#
root@k8s-master:~# showmount -e
Export list for k8s-master:
/nfsdata *
```

图 9-8

下面创建一个 PV mypv1，配置文件 nfs-pv1.yml 如图 9-9 所示。

```
apiVersion: v1
kind: PersistentVolume
metadata:
  name: mypv1
spec:
  capacity:
    storage: 1Gi          ①
  accessModes:
    - ReadWriteOnce       ②
  persistentVolumeReclaimPolicy: Recycle   ③
  storageClassName: nfs   ④
  nfs:
    path: /nfsdata/pv1
    server: 192.168.56.105               ⑤
```

图 9-9

① capacity 指定 PV 的容量为 1GB。

② accessModes 指定访问模式为 ReadWriteOnce，支持的访问模式有 3 种：ReadWriteOnce 表示 PV 能以 read-write 模式 mount 到单个节点，ReadOnlyMany 表示 PV 能以 read-only 模式 mount 到多个节点，ReadWriteMany 表示 PV 能以 read-write 模式 mount 到多个节点。

③ persistentVolumeReclaimPolicy 指定当 PV 的回收策略为 Recycle，支持的策略有 3 种：Retain 表示需要管理员手工回收；Recycle 表示清除 PV 中的数据，效果相当于执行 rm -rf /thevolume/*；Delete 表示删除 Storage Provider 上的对应存储资源，例如 AWS EBS、GCE PD、Azure Disk、OpenStack Cinder Volume 等。

④ storageClassName 指定 PV 的 class 为 nfs。相当于为 PV 设置了一个分类，PVC 可以指定 class 申请相应 class 的 PV。

⑤ 指定 PV 在 NFS 服务器上对应的目录。

创建 mypv1，如图 9-10 所示。

```
ubuntu@k8s-master:~$
ubuntu@k8s-master:~$ kubectl apply -f nfs-pv1.yml
persistentvolume "mypv1" created
ubuntu@k8s-master:~$
ubuntu@k8s-master:~$ kubectl get pv
NAME    CAPACITY   ACCESSMODES   RECLAIMPOLICY   STATUS      CLAIM   STORAGECLASS   REASON   AGE
mypv1   1Gi        RWO           Recycle         Available           nfs                     15s
ubuntu@k8s-master:~$
```

图 9-10

STATUS 为 Available，表示 mypv1 就绪，可以被 PVC 申请。

接下来创建 PVC mypvc1，配置文件 nfs-pvc1.yml 如图 9-11 所示。

```
kind: PersistentVolumeClaim
apiVersion: v1
metadata:
  name: mypvc1
spec:
  accessModes:
    - ReadWriteOnce
  resources:
    requests:
      storage: 1Gi
  storageClassName: nfs
```

图 9-11

PVC 就很简单了，只需要指定 PV 的容量、访问模式和 class 即可。

创建 mypvc1，如图 9-12 所示。

```
ubuntu@k8s-master:~$
ubuntu@k8s-master:~$ kubectl apply -f nfs-pvc1.yml
persistentvolumeclaim "mypvc1" created
ubuntu@k8s-master:~$
ubuntu@k8s-master:~$ kubectl get pvc
NAME     STATUS   VOLUME   CAPACITY   ACCESSMODES   STORAGECLASS   AGE
mypvc1   Bound    mypv1    1Gi        RWO           nfs            22s
ubuntu@k8s-master:~$
ubuntu@k8s-master:~$ kubectl get pv
NAME    CAPACITY   ACCESSMODES   RECLAIMPOLICY   STATUS   CLAIM            STORAGECLASS   REASON   AGE
mypv1   1Gi        RWO           Recycle         Bound    default/mypvc1   nfs                     6m
ubuntu@k8s-master:~$
```

图 9-12

从 kubectl get pvc 和 kubectl get pv 的输出可以看到 mypvc1 已经 Bound 到 mypv1，申请成功。

接下来就可以在 Pod 中使用存储了，Pod 配置文件 pod1.yml 如图 9-13 所示。

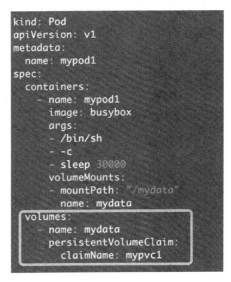

图 9-13

与使用普通 Volume 的格式类似，在 volumes 中通过 persistentVolumeClaim 指定使用 mypvc1 申请的 Volume。

创建 mypod1，如图 9-14 所示。

```
ubuntu@k8s-master:~$
ubuntu@k8s-master:~$ kubectl apply -f pod1.yml
pod "mypod1" created
ubuntu@k8s-master:~$
ubuntu@k8s-master:~$ kubectl get pod -o wide
NAME       READY    STATUS     RESTARTS    AGE      IP            NODE
mypod1     1/1      Running    0           32s      10.244.4.60   k8s-node1
ubuntu@k8s-master:~$
```

图 9-14

验证 PV 是否可用，如图 9-15 所示。

```
ubuntu@k8s-master:~$
ubuntu@k8s-master:~$ kubectl exec mypod1 touch /mydata/hello
ubuntu@k8s-master:~$
ubuntu@k8s-master:~$ ls /nfsdata/pv1/
hello
ubuntu@k8s-master:~$
```

图 9-15

可见，在 Pod 中创建的文件 /mydata/hello 确实已经保存到了 NFS 服务器目录 /nfsdata/pv1 中。

9.2.2 回收 PV

当不需要使用 PV 时，可用删除 PVC 回收 PV，如图 9-16 所示。

```
ubuntu@k8s-master:~$
ubuntu@k8s-master:~$ kubectl delete pvc mypvc1
persistentvolumeclaim "mypvc1" deleted
ubuntu@k8s-master:~$
ubuntu@k8s-master:~$ kubectl get pod -o wide
NAME                  READY   STATUS             RESTARTS   AGE    IP            NODE
mypod1                1/1     Running            0          25m    10.244.4.60   k8s-node1
recycler-for-mypv1    0/1     ContainerCreating  0          2s     <none>        k8s-node1
ubuntu@k8s-master:~$
ubuntu@k8s-master:~$ kubectl get pv
NAME    CAPACITY   ACCESSMODES   RECLAIMPOLICY   STATUS     CLAIM            STORAGECLASS   REA
mypv1   1Gi        RWO           Recycle         Released   default/mypvc1   nfs
ubuntu@k8s-master:~$
```

图 9-16

当 PVC mypvc1 被删除后，我们发现 Kubernetes 启动了一个新 Pod recycler-for-mypv1，这个 Pod 的作用就是清除 PV mypv1 的数据。此时 mypv1 的状态为 Released，表示已经解除了与 mypvc1 的 Bound，正在清除数据，不过此时还不可用。

当数据清除完毕，mypv1 的状态重新变为 Available，此时可以被新的 PVC 申请，如图 9-17 所示。

```
ubuntu@k8s-master:~$
ubuntu@k8s-master:~$ kubectl get pv
NAME     CAPACITY   ACCESSMODES   RECLAIMPOLICY   STATUS      CLAIM   STORAGECLASS
mypv1    1Gi        RWO           Recycle         Available           nfs
ubuntu@k8s-master:~$
ubuntu@k8s-master:~$ ls /nfsdata/pv1/
ubuntu@k8s-master:~$
```

图 9-17

/nfsdata/pv1 中的 hello 文件已经被删除了。

因为 PV 的回收策略设置为 Recycle，所以数据会被清除，但这可能不是我们想要的结果。如果我们希望保留数据，可以将策略设置为 Retain，如图 9-18 所示。

```
apiVersion: v1
kind: PersistentVolume
metadata:
  name: mypv1
spec:
  capacity:
    storage: 1Gi
  accessModes:
    - ReadWriteOnce
  persistentVolumeReclaimPolicy: Retain
  storageClassName: nfs
  nfs:
    path: /nfsdata/pv1
    server: 192.168.56.105
```

图 9-18

通过 kubectl apply 更新 PV，如图 9-19 所示。

```
ubuntu@k8s-master:~$
ubuntu@k8s-master:~$ kubectl apply -f nfs-pv1.yml
persistentvolume "mypv1" configured
ubuntu@k8s-master:~$
ubuntu@k8s-master:~$ kubectl get pv
NAME     CAPACITY   ACCESSMODES   RECLAIMPOLICY   STATUS      CLAIM   STORAGECLASS
mypv1    1Gi        RWO           Retain          Available           nfs
ubuntu@k8s-master:~$
```

图 9-19

回收策略已经变为 Retain，通过下面的步骤验证其效果，如图 9-20 所示。

```
ubuntu@k8s-master:~$
ubuntu@k8s-master:~$ kubectl apply -f nfs-pvc1.yml    ①
persistentvolumeclaim "mypvc1" created
ubuntu@k8s-master:~$
ubuntu@k8s-master:~$ kubectl exec mypod1 touch /mydata/hello    ②
ubuntu@k8s-master:~$
ubuntu@k8s-master:~$ ls /nfsdata/pv1/
hello
ubuntu@k8s-master:~$
ubuntu@k8s-master:~$ kubectl delete pvc mypvc1    ③
persistentvolumeclaim "mypvc1" deleted
ubuntu@k8s-master:~$
ubuntu@k8s-master:~$ kubectl get pv
NAME     CAPACITY   ACCESSMODES   RECLAIMPOLICY   STATUS     CLAIM            STORAGECLASS
mypv1    1Gi        RWO           Retain          Released   default/mypvc1   nfs
ubuntu@k8s-master:~$
ubuntu@k8s-master:~$ kubectl get pod -o wide    ④
NAME     READY   STATUS    RESTARTS   AGE   IP            NODE
mypod1   1/1     Running   0          46m   10.244.4.60   k8s-node1
ubuntu@k8s-master:~$
ubuntu@k8s-master:~$ ls /nfsdata/pv1/    ⑤
hello
ubuntu@k8s-master:~$
```

图 9-20

① 重新创建 mypvc1。
② 在 mypv1 中创建文件 hello。
③ mypv1 状态变为 Released。
④ Kubernetes 并没有启动 Pod recycler-for-mypv1。
⑤ PV 中的数据被完整保留。

虽然 mypv1 中的数据得到了保留，但其 PV 状态会一直处于 Released，不能被其他 PVC 申请。为了重新使用存储资源，可以删除并重新创建 mypv1。删除操作只是删除了 PV 对象，存储空间中的数据并不会被删除。

新建的 mypv1 状态为 Available，如图 9-21 所示，已经可以被 PVC 申请。

```
ubuntu@k8s-master:~$
ubuntu@k8s-master:~$ kubectl delete pv mypv1
persistentvolume "mypv1" deleted
ubuntu@k8s-master:~$
ubuntu@k8s-master:~$ kubectl apply -f nfs-pv1.yml
persistentvolume "mypv1" created
ubuntu@k8s-master:~$
ubuntu@k8s-master:~$ kubectl get pv
NAME    CAPACITY  ACCESSMODES  RECLAIMPOLICY  STATUS     CLAIM  STORAGECLASS  REASON  AGE
mypv1   1Gi       RWO          Retain         Available         nfs                   9s
ubuntu@k8s-master:~$
```

图 9-21

PV 还支持 Delete 的回收策略，会删除 PV 在 Storage Provider 上对应的存储空间。NFS 的 PV 不支持 Delete，支持 Delete 的 Provider 有 AWS EBS、GCE PD、Azure Disk、OpenStack Cinder Volume 等。

9.2.3 PV 动态供给

在前面的例子中，我们提前创建了 PV，然后通过 PVC 申请 PV 并在 Pod 中使用，这种方式叫作静态供给（Static Provision）。

与之对应的是动态供给（Dynamical Provision），即如果没有满足 PVC 条件的 PV，会动态创建 PV。相比静态供给，动态供给有明显的优势：不需要提前创建 PV，减少了管理员的工作量，效率高。

动态供给是通过 StorageClass 实现的，StorageClass 定义了如何创建 PV，下面给出两个例子。

（1）StorageClass standard，如图 9-22 所示。

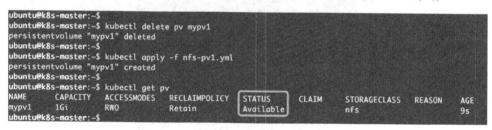

图 9-22

（2）StorageClass slow，如图 9-23 所示。

```
kind: StorageClass
apiVersion: storage.k8s.io/v1
metadata:
  name: slow
provisioner: kubernetes.io/aws-ebs
parameters:
  type: io1
  zones: us-east-1d, us-east-1c
  iopsPerGB: "10"
```

图 9-23

这两个 StorageClass 都会动态创建 AWS EBS，不同点在于 standard 创建的是 gp2 类型的 EBS，而 slow 创建的是 io1 类型的 EBS。不同类型的 EBS 支持的参数可参考 AWS 官方文档。

StorageClass 支持 Delete 和 Retain 两种 reclaimPolicy，默认是 Delete。

与之前一样，PVC 在申请 PV 时，只需要指定 StorageClass、容量以及访问模式即可，如图 9-24 所示。

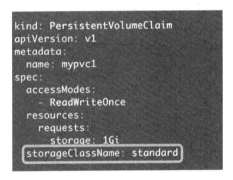

图 9-24

除了 AWS EBS，Kubernetes 还支持其他多种动态供给 PV 的 Provisioner，完整列表请参考 https://kubernetes.io/docs/concepts/storage/storage-classes/#provisioner。

9.3 一个数据库例子

本节演示如何为 MySQL 数据库提供持久化存储，步骤为：

（1）创建 PV 和 PVC。
（2）部署 MySQL。
（3）向 MySQL 添加数据。
（4）模拟节点宕机故障，Kubernetes 将 MySQL 自动迁移到其他节点。

（5）验证数据一致性。

首先创建 PV 和 PVC，配置说明如下。

- mysql-pv.yml 如图 9-25 所示。
- mysql-pvc.yml 如图 9-26 所示。

```
apiVersion: v1
kind: PersistentVolume
metadata:
  name: mysql-pv
spec:
  accessModes:
    - ReadWriteOnce
  capacity:
    storage: 1Gi
  persistentVolumeReclaimPolicy: Retain
  storageClassName: nfs
  nfs:
    path: /nfsdata/mysql-pv
    server: 192.168.56.105
```

图 9-25

```
kind: PersistentVolumeClaim
apiVersion: v1
metadata:
  name: mysql-pvc
spec:
  accessModes:
    - ReadWriteOnce
  resources:
    requests:
      storage: 1Gi
  storageClassName: nfs
```

图 9-26

创建 mysql-pv 和 mysql-pvc，如图 9-27 所示。

```
ubuntu@k8s-master:~$
ubuntu@k8s-master:~$ kubectl apply -f mysql-pv.yml
persistentvolume "mysql-pv" created
ubuntu@k8s-master:~$
ubuntu@k8s-master:~$ kubectl apply -f mysql-pvc.yml
persistentvolumeclaim "mysql-pvc" created
ubuntu@k8s-master:~$
ubuntu@k8s-master:~$ kubectl get pv,pvc
NAME          CAPACITY   ACCESSMODES   RECLAIMPOLICY   STATUS   CLAIM               STORAGECLASS
pv/mysql-pv   1Gi        RWO           Retain          Bound    default/mysql-pvc   nfs

NAME             STATUS   VOLUME     CAPACITY   ACCESSMODES   STORAGECLASS   AGE
pvc/mysql-pvc    Bound    mysql-pv   1Gi        RWO           nfs            9s
ubuntu@k8s-master:~$
```

图 9-27

接下来部署 MySQL，配置文件如图 9-28 所示。

```yaml
apiVersion: v1
kind: Service
metadata:
  name: mysql
spec:
  ports:
  - port: 3306
  selector:
    app: mysql

---
apiVersion: apps/v1beta1
kind: Deployment
metadata:
  name: mysql
spec:
  selector:
    matchLabels:
      app: mysql
  template:
    metadata:
      labels:
        app: mysql
    spec:
      containers:
      - image: mysql:5.6
        name: mysql
        env:
        - name: MYSQL_ROOT_PASSWORD
          value: password
        ports:
        - containerPort: 3306
          name: mysql
        volumeMounts:
        - name: mysql-persistent-storage
          mountPath: /var/lib/mysql
      volumes:
      - name: mysql-persistent-storage
        persistentVolumeClaim:
          claimName: mysql-pvc
```

图 9-28

PVC mysql-pvc Bound 的 PV mysql-pv 将被 mount 到 MySQL 的数据目录 var/lib/mysql，如图 9-29 所示。

```
ubuntu@k8s-master:~$
ubuntu@k8s-master:~$ kubectl apply -f mysql.yml
service "mysql" created
deployment "mysql" created
ubuntu@k8s-master:~$
ubuntu@k8s-master:~$ kubectl get pod -o wide
NAME                       READY   STATUS    RESTARTS   AGE   IP            NODE
mysql-2150355289-p99h8     1/1     Running   0          15s   10.244.5.80   k8s-node2
ubuntu@k8s-master:~$
```

图 9-29

MySQL 被部署到 k8s-node2，下面通过客户端访问 Service mysql，如图 9-30 所示。

```
kubectl run -it --rm --image=mysql:5.6 --restart=Never mysql-client -- mysql -h mysql -ppassword
```

```
ubuntu@k8s-master:~$
ubuntu@k8s-master:~$ kubectl run -it --rm --image=mysql:5.6 --restart=Never mysql-client -- mysql -h mysql -p
If you don't see a command prompt, try pressing enter.

mysql>
```

图 9-30

更新数据库，如图 9-31 所示。

```
mysql>
mysql> use mysql    ①
Reading table information for completion of table and column names
You can turn off this feature to get a quicker startup with -A

Database changed
mysql> create table my_id( id int(4) );   ②
Query OK, 0 rows affected (0.01 sec)

mysql> insert my_id values( 111 );   ③
Query OK, 1 row affected (0.00 sec)

mysql> select * from my_id;   ④
+------+
| id   |
+------+
| 111  |
+------+
1 row in set (0.00 sec)

mysql>
```

图 9-31

① 切换到数据库 mysql。
② 创建数据库表 my_id。
③ 插入一条数据。
④ 确认数据已经写入。

关闭 k8s-node2，模拟节点宕机故障，如图 9-32 所示。

```
root@k8s-node2:~#
root@k8s-node2:~# shutdown now
Connection to 192.168.56.107 closed by remote host.
Connection to 192.168.56.107 closed.
```

图 9-32

一段时间后，Kubernetes 将 MySQL 迁移到 k8s-node1，如图 9-33 所示。

```
ubuntu@k8s-master:~$
ubuntu@k8s-master:~$ kubectl get pod -o wide
NAME                     READY   STATUS    RESTARTS   AGE   IP            NODE
mysql-2150355289-p13n5   1/1     Running   0          31s   10.244.4.66   k8s-node1
mysql-2150355289-p99h8   1/1     Unknown   0          16m   10.244.5.80   k8s-node2
ubuntu@k8s-master:~$
```

图 9-33

验证数据的一致性，如图 9-34 所示。

```
ubuntu@k8s-master:~$
ubuntu@k8s-master:~$ kubectl run -it --rm --image=mysql:5.6 --restart=Never mysql-client -- mysql -h mysql -ppassword
If you don't see a command prompt, try pressing enter.

mysql> use mysql
Reading table information for completion of table and column names
You can turn off this feature to get a quicker startup with -A

Database changed
mysql> select * from my_id;
+------+
| id   |
+------+
| 111  |
+------+
1 row in set (0.00 sec)

mysql>
```

图 9-34

MySQL 服务恢复，数据也完好无损。

9.4 小结

本章我们讨论了 Kubernetes 如何管理存储资源。

emptyDir 和 hostPath 类型的 Volume 很方便，但可持久性不强，Kubernetes 支持多种外部存储系统的 Volume。

PV 和 PVC 分离了管理员和普通用户的职责，更适合生产环境。我们还学习了如何通过 StorageClass 实现更高效的动态供给。

最后，我们演示了如何在 MySQL 中使用 PersistentVolume 实现数据持久性。

第 10 章
Secret & Configmap

应用启动过程中可能需要一些敏感信息，比如访问数据库的用户名、密码或者密钥。将这些信息直接保存在容器镜像中显然不妥，Kubernetes 提供的解决方案是 Secret。

Secret 会以密文的方式存储数据，避免了直接在配置文件中保存敏感信息。Secret 会以 Volume 的形式被 mount 到 Pod，容器可通过文件的方式使用 Secret 中的敏感数据；此外，容器也可以环境变量的方式使用这些数据。

Secret 可通过命令行或 YAML 创建。比如希望 Secret 中包含如下信息：用户名 admin、密码 123456。

10.1 创建 Secret

有四种方法创建 Secret：

（1）通过 --from-literal：

```
kubectl create secret generic mysecret --from-literal=username=admin --from-literal=password=123456
```

每个 --from-literal 对应一个信息条目。

（2）通过 --from-file：

```
echo -n admin > ./username
echo -n 123456 > ./password
kubectl create secret generic mysecret --from-file=./username --from-file=./password
```

每个文件内容对应一个信息条目。

（3）通过 --from-env-file：

```
cat << EOF > env.txt
username=admin
password=123456
EOF
kubectl create secret generic mysecret --from-env-file=env.txt
```

文件 env.txt 中每行 Key=Value 对应一个信息条目。

（4）通过 YAML 配置文件，如图 10-1 所示。

图 10-1

文件中的敏感数据必须是通过 base64 编码后的结果，如图 10-2 所示。

图 10-2

执行 kubectl apply 创建 Secret，如图 10-3 所示。

图 10-3

10.2 查看 Secret

通过 kubectl get secret 查看存在的 secret，如图 10-4 所示。

图 10-4

显示有两个数据条目，通过 kubectl describe secret 查看条目的 Key，如图 10-5 所示。

```
ubuntu@k8s-master:~$
ubuntu@k8s-master:~$ kubectl describe secret mysecret
Name:           mysecret
Namespace:      default
Labels:         <none>
Annotations:    <none>

Type:   Opaque

Data
====
password:    6 bytes
username:    5 bytes
ubuntu@k8s-master:~$
```

图 10-5

如果还想查看 Value，可以用 kubectl edit secret mysecret，如图 10-6 所示。

```
apiVersion: v1
data:
  password: MTIzNDU2
  username: YWRtaW4=
kind: Secret
metadata:
  creationTimestamp: 2017-10-10T07:16:21Z
  name: mysecret
  namespace: default
  resourceVersion: "1598871"
  selfLink: /api/v1/namespaces/default/secrets/mysecret
  uid: eaeb4980-ad8a-11e7-8f72-0800274451ad
type: Opaque
```

图 10-6

然后通过 base64 将 Value 反编码，如图 10-7 所示。

```
ubuntu@k8s-master:~$
ubuntu@k8s-master:~$ echo -n MTIzNDU2 | base64 --decode
123456ubuntu@k8s-master:~$
ubuntu@k8s-master:~$ echo -n YWRtaW4= | base64 --decode
adminubuntu@k8s-master:~$
ubuntu@k8s-master:~$
```

图 10-7

10.3 在 Pod 中使用 Secret

Pod 可以通过 Volume 或者环境变量的方式使用 Secret。

10.3.1 Volume 方式

Pod 的配置文件如图 10-8 所示。

① 定义 volume foo，来源为 secret mysecret。
② 将 foo mount 到容器路径 /etc/foo，可指定读写权限为 readOnly。

创建 Pod 并在容器中读取 Secret，如图 10-9 所示。

```
apiVersion: v1
kind: Pod
metadata:
  name: mypod
spec:
  containers:
  - name: mypod
    image: busybox
    args:
      - /bin/sh
      - -c
      - sleep 10; touch /tmp/healthy; sleep 30000
    volumeMounts:
    - name: foo
      mountPath: "/etc/foo"   ②
      readOnly: true
  volumes:
  - name: foo   ①
    secret:
      secretName: mysecret
```

图 10-8

```
ubuntu@k8s-master:~$
ubuntu@k8s-master:~$ kubectl apply -f mypod.yml
pod "mypod" created
ubuntu@k8s-master:~$
ubuntu@k8s-master:~$ kubectl exec -it mypod sh
/ #
/ # ls /etc/foo
password   username
/ #
/ # cat /etc/foo/username
admin/ #
/ # cat /etc/foo/password
123456/ #
/ #
```

图 10-9

可以看到，Kubernetes 会在指定的路径 /etc/foo 下为每条敏感数据创建一个文件，文件名就是数据条目的 Key，这里是 /etc/foo/username 和 /etc/foo/password，Value 则以明文存放在文件中。

我们也可以自定义存放数据的文件名，比如将配置文件改为如图 10-10 所示那样。

```
apiVersion: v1
kind: Pod
metadata:
  name: mypod
spec:
  containers:
  - name: mypod
    image: busybox
    args:
      - /bin/sh
      - -c
      - sleep 10; touch /tmp/healthy; sleep 30000
    volumeMounts:
    - name: foo
      mountPath: "/etc/foo"
      readOnly: true
  volumes:
  - name: foo
    secret:
      secretName: mysecret
      items:
      - key: username
        path: my-group/my-username
      - key: password
        path: my-group/my-password
```

图 10-10

这时数据将分别存放在/etc/foo/my-group/my-username 和/etc/foo/my-group/my-password 中。

以 Volume 方式使用的 Secret 支持动态更新：Secret 更新后，容器中的数据也会更新。

将 password 更新为 abcdef，base64 编码为 YWJjZGVm，如图 10-11 所示。

更新 Secret，如图 10-12 所示。

```
apiVersion: v1
kind: Secret
metadata:
  name: mysecret
data:
  username: YWRtaW4=
  password: YWJjZGVm
```

图 10-11

```
ubuntu@k8s-master:~$
ubuntu@k8s-master:~$ kubectl apply -f mysecrete.yml
secret "mysecret" configured
ubuntu@k8s-master:~$
```

图 10-12

几秒钟后，新的 password 会同步到容器，如图 10-13 所示。

```
/ #
/ # cat /etc/foo/password
123456/ #
/ #
/ # cat /etc/foo/password
abcdef/ #
/ #
```

图 10-13

10.3.2 环境变量方式

通过 Volume 使用 Secret，容器必须从文件读取数据，稍显麻烦，Kubernetes 还支持通过环境变量使用 Secret。

Pod 配置文件示例如图 10-14 所示。

```
apiVersion: v1
kind: Pod
metadata:
  name: mypod
spec:
  containers:
  - name: mypod
    image: busybox
    args:
      - /bin/sh
      - -c
      - sleep 10; touch /tmp/healthy; sleep 30000
    env:
      - name: SECRET_USERNAME
        valueFrom:
          secretKeyRef:
            name: mysecret
            key: username
      - name: SECRET_PASSWORD
        valueFrom:
          secretKeyRef:
            name: mysecret
            key: password
```

图 10-14

创建 Pod 并读取 Secret，如图 10-15 所示。

```
ubuntu@k8s-master:~$
ubuntu@k8s-master:~$ kubectl apply -f mypod-env.yml
pod "mypod" created
ubuntu@k8s-master:~$
ubuntu@k8s-master:~$ kubectl exec -it mypod sh
/ #
/ # echo $SECRET_USERNAME
admin
/ #
/ # echo $SECRET_PASSWORD
abcdef
/ #
```

图 10-15

通过环境变量 SECRET_USERNAME 和 SECRET_PASSWORD 成功读取到 Secret 的数据。

需要注意的是，环境变量读取 Secret 很方便，但无法支撑 Secret 动态更新。

10.4 ConfigMap

Secret 可以为 Pod 提供密码、Token、私钥等敏感数据；对于一些非敏感数据，比如应用的配置信息，则可以用 ConfigMap。

ConfigMap 的创建和使用方式与 Secret 非常类似，主要的不同是数据以明文的形式存放。

与 Secret 一样，ConfigMap 也支持四种创建方式：

（1）通过 --from-literal：

```
kubectl create configmap myconfigmap --from-literal=config1=xxx --from-literal=config2=yyy
```

每个 --from-literal 对应一个信息条目。

（2）通过 --from-file：

```
echo -n xxx > ./config1
echo -n yyy > ./config2
kubectl create configmap myconfigmap --from-file=./config1 --from-file=./config2
```

每个文件内容对应一个信息条目。

（3）通过 --from-env-file：

```
cat << EOF > env.txt
config1=xxx
config2=yyy
EOF
kubectl create configmap myconfigmap --from-env-file=env.txt
```

文件 env.txt 中每行 Key=Value 对应一个信息条目。

（4）通过 YAML 配置文件，如图 10-16 所示。文件中的数据直接以明文输入。

与 Secret 一样，Pod 也可以通过 Volume 或者环境变量的方式使用 Secret。

（1）Volume 方式如图 10-17 所示。

```
apiVersion: v1
kind: ConfigMap
metadata:
  name: myconfigmap
data:
  config1: xxx
  config2: yyy
```

图 10-16

```
apiVersion: v1
kind: Pod
metadata:
  name: mypod
spec:
  containers:
  - name: mypod
    image: busybox
    args:
      - /bin/sh
      - -c
      - sleep 10; touch /tmp/healthy; sleep 30000
    volumeMounts:
    - name: foo
      mountPath: "/etc/foo"
      readOnly: true
  volumes:
  - name: foo
    configMap:
      name: myconfigmap
```

图 10-17

（2）环境变量方式如图 10-18 所示。

```
apiVersion: v1
kind: Pod
metadata:
  name: mypod
spec:
  containers:
  - name: mypod
    image: busybox
    args:
      - /bin/sh
      - -c
      - sleep 10; touch /tmp/healthy; sleep 30000
    env:
    - name: CONFIG_1
      valueFrom:
        configMapKeyRef:
          name: myconfigmap
          key: config1
    - name: CONFIG_2
      valueFrom:
        configMapKeyRef:
          name: myconfigmap
          key: config2
```

图 10-18

大多数情况下，配置信息都以文件形式提供，所以在创建 ConfigMap 时通常采用 --from-file 或 YAML 方式，读取 ConfigMap 时通常采用 Volume 方式。比如给 Pod 传递如何记录日志的配置信息，如图 10-19 所示。

```
class: logging.handlers.RotatingFileHandler
formatter: precise
level: INFO
filename: %hostname-%timestamp.log
```

图 10-19

可以采用 --from-file 形式，将其保存在文件 logging.conf 中，然后执行命令：

```
kubectl create configmap myconfigmap --from-file=./logging.conf
```

如果采用 YAML 配置文件，其内容则如图 10-20 所示。

```
apiVersion: v1
kind: ConfigMap
metadata:
  name: myconfigmap
data:
  logging.conf: |
    class: logging.handlers.RotatingFileHandler
    formatter: precise
    level: INFO
    filename: %hostname-%timestamp.log
```

图 10-20

注意，别漏写了 Key logging.conf 后面的 | 符号。

创建并查看 ConfigMap，如图 10-21 所示。

```
ubuntu@k8s-master:~$
ubuntu@k8s-master:~$ kubectl apply -f myconfigmap.yml
configmap "myconfigmap" created
ubuntu@k8s-master:~$
ubuntu@k8s-master:~$ kubectl get configmap myconfigmap
NAME          DATA      AGE
myconfigmap   1         12s
ubuntu@k8s-master:~$
ubuntu@k8s-master:~$ kubectl describe configmap myconfigmap
Name:         myconfigmap
Namespace:    default
Labels:       <none>
Annotations:  kubectl.kubernetes.io/last-applied-configuration={"apiVersion":"v1"
e-%timestamp...

Data
====
logging.conf:
----
class: logging.handlers.RotatingFileHandler
formatter: precise
level: INFO
filename: %hostname-%timestamp.log

Events: <none>
ubuntu@k8s-master:~$
```

图 10-21

在 Pod 中使用此 ConfigMap，配置文件如图 10-22 所示。

```
apiVersion: v1
kind: Pod
metadata:
  name: mypod
spec:
  containers:
  - name: mypod
    image: busybox
    args:
      - /bin/sh
      - -c
      - sleep 10; touch /tmp/healthy; sleep 30000
    volumeMounts:
    - name: foo
      mountPath: "/etc"          ②
  volumes:
  - name: foo
    configMap:
      name: myconfigmap
      items:
        - key: logging.conf       ①
          path: myapp/logging.conf
```

图 10-22

① 在 volume 中指定存放配置信息的文件相对路径为 myapp/logging.conf。
② 将 volume mount 到容器的 /etc 目录。

创建 Pod 并读取配置信息，如图 10-23 所示。

```
ubuntu@k8s-master:~$
ubuntu@k8s-master:~$ kubectl apply -f mypod.yml
pod "mypod" created
ubuntu@k8s-master:~$
ubuntu@k8s-master:~$ kubectl exec -it mypod sh
/ #
/ # cat /etc/myapp/logging.conf
class: logging.handlers.RotatingFileHandler
formatter: precise
level: INFO
filename: %hostname-%timestamp.log
/ #
```

图 10-23

配置信息已经保存到 /etc/myapp/logging.conf 文件中。与 Secret 一样，Volume 形式的 ConfigMap 也支持动态更新，留给大家自己实践。

10.5 小结

本章我们学习了如何向 Pod 传递配置信息。如果信息需要加密，可使用 Secret；如果是一般的配置信息，则可使用 ConfigMap。

Secret 和 ConfigMap 支持四种定义方法。Pod 在使用它们时，可以选择 Volume 方式或环境变量方式，不过只有 Volume 方式支持动态更新。

第 11 章

Helm——Kubernetes 的包管理器

本章我们将学习 Helm —— Kubernetes 的包管理器。

每个成功的软件平台都有一个优秀的打包系统，比如 Debian、Ubuntu 的 apt，Red Hat、CentOS 的 yum。Helm 则是 Kubernetes 上的包管理器。

本章我们将讨论为什么需要 Helm、它的架构和组件，以及如何使用 Helm。

11.1 Why Helm

Helm 到底解决了什么问题？为什么 Kubernetes 需要 Helm？

答案是：Kubernetes 能够很好地组织和编排容器，但它缺少一个更高层次的应用打包工具，而 Helm 就是来干这件事的。

先来看个例子。

比如对于一个 MySQL 服务，Kubernetes 需要部署下面这些对象：

（1）Service，让外界能够访问到 MySQL，如图 11-1 所示。

```
apiVersion: v1
kind: Service
metadata:
  name: my-mysql
  labels:
    app: my-mysql
spec:
  ports:
  - name: mysql
    port: 3306
    targetPort: mysql
  selector:
    app: my-mysql
```

图 11-1

（2）Secret，定义 MySQL 的密码，如图 11-2 所示。

（3）PersistentVolumeClaim，为 MySQL 申请持久化存储空间，如图 11-3 所示。

```
apiVersion: v1
kind: Secret
metadata:
  name: my-mysql
  labels:
    app: my-mysql
type: Opaque
data:
  mysql-root-password: "M0MzREhRQWRjeQ=="
  mysql-password: "eGNXZkpMNmlkSw=="
```

图 11-2

```
kind: PersistentVolumeClaim
apiVersion: v1
metadata:
  name: my-mysql
  labels:
    app: my-mysql
spec:
  accessModes:
    - "ReadWriteOnce"
  resources:
    requests:
      storage: "8Gi"
```

图 11-3

（4）Deployment，部署 MySQL Pod，并使用上面的这些支持对象，如图 11-4 所示。

```
apiVersion: extensions/v1beta1
kind: Deployment
metadata:
  name: my-mysql
  labels:
    app: my-mysql
spec:
  template:
    metadata:
      labels:
        app: my-mysql
    spec:
      containers:
      - name: my-mysql
        image: "mysql:5.7.14"
        env:
        - name: MYSQL_ROOT_PASSWORD
          valueFrom:
            secretKeyRef:
              name: my-mysql
              key: mysql-root-password
        - name: MYSQL_PASSWORD
          valueFrom:
            secretKeyRef:
              name: my-mysql
              key: mysql-password
        - name: MYSQL_USER
          value: ""
        - name: MYSQL_DATABASE
          value: ""
        ports:
        - name: mysql
          containerPort: 3306
        volumeMounts:
        - name: data
          mountPath: /var/lib/mysql
      volumes:
      - name: data
        persistentVolumeClaim:
          claimName: my-mysql
```

图 11-4

我们可以将上面这些配置保存到对象各自的文件中，或者集中写进一个配置文件，然后通过 kubectl apply -f 部署。

到目前为止，Kubernetes 对服务的部署支持得都挺好，如果应用只由一个或几个这样的服务组成，上面的部署方式完全足够了。

但是，如果我们开发的是微服务架构的应用，组成应用的服务可能多达十个甚至几十上百个，这种组织和管理应用的方式就不好使了：

（1）很难管理、编辑和维护如此多的服务。每个服务都有若干配置，缺乏一个更高层次的工具将这些配置组织起来。

（2）不容易将这些服务作为一个整体统一发布。部署人员需要首先理解应用都包含哪些服务，然后按照逻辑顺序依次执行 kubectl apply，即缺少一种工具来定义应用与服务，以及服务与服务之间的依赖关系。

（3）不能高效地共享和重用服务。比如两个应用都要用到 MySQL 服务，但配置的参数不一样，这两个应用只能分别复制一套标准的 MySQL 配置文件，修改后通过 kubectl apply 部署。也就是说，不支持参数化配置和多环境部署。

（4）不支持应用级别的版本管理。虽然可以通过 kubectl rollout undo 进行回滚，但这只能针对单个 Deployment，不支持整个应用的回滚。

（5）不支持对部署的应用状态进行验证。比如是否能通过预定义的账号访问 MySQL。虽然 Kubernetes 有健康检查，但那是针对单个容器，我们需要应用（服务）级别的健康检查。

Helm 能够解决上面这些问题，Helm 帮助 Kubernetes 成为微服务架构应用理想的部署平台。

11.2 Helm 架构

在实践之前，我们先来看看 Helm 的架构。

Helm 有两个重要的概念：chart 和 release。

- chart 是创建一个应用的信息集合，包括各种 Kubernetes 对象的配置模板、参数定义、依赖关系、文档说明等。chart 是应用部署的自包含逻辑单元。可以将 chart 想象成 apt、yum 中的软件安装包。
- release 是 chart 的运行实例，代表了一个正在运行的应用。当 chart 被安装到 Kubernetes 集群，就生成一个 release。chart 能够多次安装到同一个集群，每次安装都是一个 release。

Helm 是包管理工具，这里的包就是指的 chart。Helm 能够：

- 从零创建新 chart。
- 与存储 chart 的仓库交互，拉取、保存和更新 chart。
- 在 Kubernetes 集群中安装和卸载 release。
- 更新、回滚和测试 release。

Helm 包含两个组件：Helm 客户端和 Tiller 服务器，如图 11-5 所示。

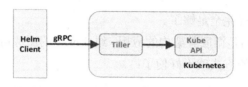

图 11-5

Helm 客户端是终端用户使用的命令行工具，用户可以：

- 在本地开发 chart。
- 管理 chart 仓库。
- 与 Tiller 服务器交互。
- 在远程 Kubernetes 集群上安装 chart。
- 查看 release 信息。
- 升级或卸载已有的 release。

Tiller 服务器运行在 Kubernetes 集群中，它会处理 Helm 客户端的请求，与 Kubernetes API Server 交互。Tiller 服务器负责：

- 监听来自 Helm 客户端的请求。
- 通过 chart 构建 release。
- 在 Kubernetes 中安装 chart，并跟踪 release 的状态。
- 通过 API Server 升级或卸载已有的 release。

简单地讲，Helm 客户端负责管理 chart，Tiller 服务器负责管理 release。

11.3 安装 Helm

本节我们将依次安装 Helm 客户端和 Tiller 服务器。

11.3.1 Helm 客户端

通常，我们将 Helm 客户端安装在能够执行 kubectl 命令的节点上，只需要下面一条命令：

```
curl https://raw.githubusercontent.com/kubernetes/helm/master/scripts/get | bash
```

结果如图 11-6 所示。

```
ubuntu@k8s-master:~$
ubuntu@k8s-master:~$ curl https://raw.githubusercontent.com/kubernetes/helm/master/scripts/get | bash
  % Total    % Received % Xferd  Average Speed   Time    Time     Time  Current
                                 Dload  Upload   Total   Spent    Left  Speed
100  6329  100  6329    0     0   4179      0  0:00:01  0:00:01 --:--:--  4180
Downloading https://kubernetes-helm.storage.googleapis.com/helm-v2.6.2-linux-amd64.tar.gz
Preparing to install into /usr/local/bin
helm installed into /usr/local/bin/helm
Run 'helm init' to configure helm.
ubuntu@k8s-master:~$
```

图 11-6

执行 helm version 验证，如图 11-7 所示。

```
ubuntu@k8s-master:~$
ubuntu@k8s-master:~$ helm version
Client: &version.Version{SemVer:"v2.7.0", GitCommit:"08c1144f5eb3e3b636
d9775617287cc26e53dba4", GitTreeState:"clean"}
Error: cannot connect to Tiller
ubuntu@k8s-master:~$
```

图 11-7

目前只能查看到客户端的版本，服务器还没有安装。

helm 有很多子命令和参数，为了提高使用命令行的效率，通常建议安装 helm 的 bash 命令补全脚本，方法如下：

```
helm completion bash > .helmrc echo "source .helmrc" >> .bashrc
```

重新登录后就可以通过 Tab 键补全 helm 子命令和参数了，如图 11-8 所示。

```
ubuntu@k8s-master:~$
ubuntu@k8s-master:~$ helm
completion   get        install    repo       status      version
create       history    lint       reset      template
delete       home       list       rollback   test
dependency   init       package    search     upgrade
fetch        inspect    plugin     serve      verify
ubuntu@k8s-master:~$ helm install --
--ca-file=              --name=              --tls-ca-cert=
--cert-file=            --namespace=         --tls-cert=
--debug                 --name-template=     --tls-key=
--devel                 --no-hooks           --tls-verify
--dry-run               --replace            --values=
--home=                 --repo=              --verify
--host=                 --set=               --version=
--key-file=             --tiller-namespace=  --wait
--keyring=              --timeout=
--kube-context=         --tls
ubuntu@k8s-master:~$
```

图 11-8

11.3.2 Tiller 服务器

Tiller 服务器安装非常简单，只需要执行 helm init 即可，如图 11-9 所示。

```
ubuntu@k8s-master:~$
ubuntu@k8s-master:~$ helm init
$HELM_HOME has been configured at /home/ubuntu/.helm.

Tiller (the Helm server-side component) has been installed into your Kubernetes Cluster.
Happy Helming!
ubuntu@k8s-master:~$
```

图 11-9

Tiller 本身也是作为容器化应用运行在 Kubernetes Cluster 中的，如图 11-10 所示。

```
ubuntu@k8s-master:~$
ubuntu@k8s-master:~$ kubectl get --namespace=kube-system svc tiller-deploy
NAME           CLUSTER-IP      EXTERNAL-IP   PORT(S)     AGE
tiller-deploy  10.98.124.71    <none>        44134/TCP   1m
ubuntu@k8s-master:~$
ubuntu@k8s-master:~$ kubectl get --namespace=kube-system deployment tiller-deploy
NAME           DESIRED   CURRENT   UP-TO-DATE   AVAILABLE   AGE
tiller-deploy  1         1         1            1           1m
ubuntu@k8s-master:~$
ubuntu@k8s-master:~$ kubectl get --namespace=kube-system pod tiller-deploy-1936853538-c8sfp
NAME                             READY   STATUS    RESTARTS   AGE
tiller-deploy-1936853538-c8sfp   1/1     Running   0          2m
ubuntu@k8s-master:~$
```

图 11-10

可以看到 Tiller 的 Service、Deployment 和 Pod。

现在，helm version 已经能够查看到服务器的版本信息了，如图 11-11 所示。

```
root@k8s-master:~# helm version
Client: &version.Version{SemVer:"v2.7.0", GitCommit:"08c1144f5eb3e3b636d9775617287cc
26e53dba4", GitTreeState:"clean"}
Server: &version.Version{SemVer:"v2.7.0", GitCommit:"08c1144f5eb3e3b636d9775617287cc
26e53dba4", GitTreeState:"clean"}
```

图 11-11

11.4 使用 Helm

Helm 安装成功后，可执行 helm search 查看当前可安装的 chart，如图 11-12 所示。

```
ubuntu@k8s-master:~$
ubuntu@k8s-master:~$ helm search
NAME                             VERSION   DESCRIPTION
local/cool-chart                 0.1.0     A Helm chart for Kubernetes
stable/acs-engine-autoscaler     2.1.0     Scales worker nodes within agent pools
stable/artifactory               6.1.0     Universal Repository Manager supporting all maj...
stable/aws-cluster-autoscaler    0.3.0     Scales worker nodes within autoscaling groups.
stable/buildkite                 0.2.0     Agent for Buildkite
stable/centrifugo                2.0.0     Centrifugo is a real-time messaging server.
stable/chaoskube                 0.5.0     Chaoskube periodically kills random pods in you...
stable/chronograf                0.3.0     Open-source web application written in Go and R...
stable/cluster-autoscaler        0.2.0     Scales worker nodes within autoscaling groups.
stable/cockroachdb               0.5.0     CockroachDB is a scalable, survivable, strongly...
stable/concourse                 0.7.0     Concourse is a simple and scalable CI system.
stable/consul                    0.4.2     Highly available and distributed service discov...
stable/coredns                   0.5.0     CoreDNS is a DNS server that chains middleware ...
stable/coscale                   0.2.0     CoScale Agent
stable/dask-distributed          2.0.0     Distributed computation in Python
stable/datadog                   0.8.0     DataDog Agent
stable/dokuwiki                  0.2.0     DokuWiki is a standards-compliant, simple to us...
stable/drupal                    0.10.2    One of the most versatile open source content m...
stable/etcd-operator             0.5.0     CoreOS etcd-operator Helm chart for Kubernetes
```

图 11-12

这个列表很长，这里只截取了一部分。大家不禁会问，这些 chart 都是从哪里来的？

前面说过，Helm 可以像 apt 和 yum 管理软件包一样管理 chart。apt 和 yum 的软件包存放在仓库中，同样的，Helm 也有仓库，如图 11-13 所示。

```
ubuntu@k8s-master:~$
ubuntu@k8s-master:~$ helm repo list
NAME      URL
stable    https://kubernetes-charts.storage.googleapis.com
local     http://127.0.0.1:8879/charts
ubuntu@k8s-master:~$
```

图 11-13

Helm 安装时已经默认配置好了两个仓库：stable 和 local。stable 是官方仓库，local 是用户存放自己开发的 chart 的本地仓库。

helm search 会显示 chart 位于哪个仓库，比如 local/cool-chart 和 stable/acs-engine-autoscaler。

用户可以通过 helm repo add 添加更多的仓库，比如企业的私有仓库，仓库的管理和维护方法请参考官网文档 https://docs.helm.sh。

与 apt 和 yum 一样，helm 也支持关键字搜索，如图 11-14 所示。包括 DESCRIPTION 在内的所有信息，只要跟关键字匹配，都会显示在结果列表中。

```
ubuntu@k8s-master:~$
ubuntu@k8s-master:~$ helm search mysql
NAME                    VERSION  DESCRIPTION
stable/mysql            0.3.0    Fast, reliable, scalable, and easy to use open-...
stable/percona          0.3.0    free, fully compatible, enhanced, open source d...
stable/gcloud-sqlproxy  0.2.0    Google Cloud SQL Proxy
stable/mariadb          2.0.0    Fast, reliable, scalable, and easy to use open-...
ubuntu@k8s-master:~$
```

图 11-14

安装 chart 也很简单，执行如下命令就可以安装 MySQL。

```
helm install stable/mysql
```

如果看到如图 11-15 所示的报错，通常是因为 Tiller 服务器的权限不足。

```
ubuntu@k8s-master:~$
ubuntu@k8s-master:~$ helm install stable/mysql
Error: no available release name found
ubuntu@k8s-master:~$
```

图 11-15

执行如下命名添加权限：

```
kubectl create serviceaccount --namespace kube-system tiller
kubectl create clusterrolebinding tiller-cluster-rule
--clusterrole=cluster-admin --serviceaccount=kube-system:tiller
kubectl patch deploy --namespace kube-system tiller-deploy -p
'{"spec":{"template":{"spec":{"serviceAccount":"tiller"}}}}'
```

然后再次执行下面的命令，结果如图 11-16 所示。

```
helm install stable/mysql
```

```
ubuntu@k8s-master:~$
ubuntu@k8s-master:~$ helm install stable/mysql
NAME:   fun-zorse  ①
LAST DEPLOYED: Tue Oct 17 11:34:55 2017
NAMESPACE: default
STATUS: DEPLOYED

RESOURCES:      ②
==> v1/Service
NAME             CLUSTER-IP      EXTERNAL-IP   PORT(S)    AGE
fun-zorse-mysql  10.108.23.5     <none>        3306/TCP   0s

==> v1beta1/Deployment
NAME             DESIRED   CURRENT   UP-TO-DATE   AVAILABLE   AGE
fun-zorse-mysql  1         1         1            0           0s

==> v1/Secret
NAME             TYPE     DATA   AGE
fun-zorse-mysql  Opaque   2      1s

==> v1/PersistentVolumeClaim
NAME             STATUS    VOLUME   CAPACITY   ACCESSMODES   STORAGECLASS   AGE
fun-zorse-mysql  Pending   1s


NOTES:   ③
MySQL can be accessed via port 3306 on the following DNS name from within your cluster:
fun-zorse-mysql.default.svc.cluster.local

To get your root password run:

    kubectl get secret --namespace default fun-zorse-mysql -o jsonpath="{.data.mysql-root-password

To connect to your database:

1. Run an Ubuntu pod that you can use as a client:

    kubectl run -i --tty ubuntu --image=ubuntu:16.04 --restart=Never -- bash -il

2. Install the mysql client:

    $ apt-get update && apt-get install mysql-client -y

3. Connect using the mysql cli, then provide your password:
    $ mysql -h fun-zorse-mysql -p
```

图 11-16

输出分为三部分：

① chart 本次部署的描述信息。

- NAME 是 release 的名字，因为我们没用 -n 参数指定，所以 Helm 随机生成了一个，这里是 fun-zorse。
- NAMESPACE 是 release 部署的 namespace，默认是 default，也可以通过 --namespace 指定。
- STATUS 为 DEPLOYED，表示已经将 chart 部署到集群。

② 当前 release 包含的资源：Service、Deployment、Secret 和 PersistentVolumeClaim，其名字都是 fun-zorse-mysql，命名的格式为 ReleasName-ChartName。

③ NOTES 部分显示的是 release 的使用方法，比如如何访问 Service、如何获取数据库密码以及如何连接数据库等。

通过 kubectl get 可以查看组成 release 的各个对象，如图 11-17 所示。

```
ubuntu@k8s-master:~$
ubuntu@k8s-master:~$ kubectl get service fun-zorse-mysql
NAME              CLUSTER-IP     EXTERNAL-IP   PORT(S)    AGE
fun-zorse-mysql   10.108.23.5    <none>        3306/TCP   15m
ubuntu@k8s-master:~$
ubuntu@k8s-master:~$ kubectl get deployment fun-zorse-mysql
NAME              DESIRED   CURRENT   UP-TO-DATE   AVAILABLE   AGE
fun-zorse-mysql   1         1         1            0           15m
ubuntu@k8s-master:~$
ubuntu@k8s-master:~$ kubectl get pod fun-zorse-mysql-2959529493-hfq1t
NAME                               READY   STATUS    RESTARTS   AGE
fun-zorse-mysql-2959529493-hfq1t   0/1     Pending   0          15m
ubuntu@k8s-master:~$
ubuntu@k8s-master:~$ kubectl get pvc fun-zorse-mysql
NAME              STATUS    VOLUME   CAPACITY   ACCESSMODES   STORAGECLASS   AGE
fun-zorse-mysql   Pending                                                    15m
ubuntu@k8s-master:~$
```

图 11-17

因为我们还没有准备 PersistentVolume，所以当前 release 还不可用。

helm list 显示已经部署的 release，helm delete 可以删除 release，如图 11-18 所示。

```
ubuntu@k8s-master:~$
ubuntu@k8s-master:~$ helm list
NAME        REVISION   UPDATED                      STATUS     CHART         NAMESPAC
fun-zorse   1          Tue Oct 17 11:34:55 2017     DEPLOYED   mysql-0.3.0   default
ubuntu@k8s-master:~$
ubuntu@k8s-master:~$ helm delete fun-zorse
release "fun-zorse" deleted
ubuntu@k8s-master:~$
```

图 11-18

Helm 的使用方法像极了 apt 和 yum，用 Helm 来管理 Kubernetes 应用非常方便。

11.5 chart 详解

chart 是 Helm 的应用打包格式。chart 由一系列文件组成，这些文件描述了 Kubernetes 部署应用时所需要的资源，比如 Service、Deployment、PersistentVolumeClaim、Secret、ConfigMap 等。

单个的 chart 可以非常简单，只用于部署一个服务，比如 Memcached。chart 也可以很复杂，部署整个应用，比如包含 HTTP Servers、Database、消息中间件、Cache 等。

chart 将这些文件放置在预定义的目录结构中，通常整个 chart 被打成 tar 包，而且标注上版本信息，便于 Helm 部署。

下面我们将详细讨论 chart 的目录结构以及包含的各类文件。

11.5.1 chart 目录结构

以前面 MySQL chart 为例。一旦安装了某个 chart，我们就可以在 ~/.helm/cache/archive 中找到 chart 的 tar 包，如图 11-19 所示。

```
ubuntu@k8s-master:~$
ubuntu@k8s-master:~$ ls ~/.helm/cache/archive
mysql-0.3.0.tgz
ubuntu@k8s-master:~$
```

图 11-19

解压后,MySQL chart 目录结构如图 11-20 所示。

```
ubuntu@k8s-master:~$
ubuntu@k8s-master:~$ tree mysql
mysql
├── Chart.yaml
├── README.md
├── templates
│   ├── configmap.yaml
│   ├── deployment.yaml
│   ├── _helpers.tpl
│   ├── NOTES.txt
│   ├── pvc.yaml
│   ├── secrets.yaml
│   └── svc.yaml
└── values.yaml

1 directory, 10 files
ubuntu@k8s-master:~$
```

图 11-20

目录名就是 chart 的名字(不带版本信息),这里是 mysql,包含如下内容。

(1) Chart.yaml

YAML 文件,描述 chart 的概要信息,如图 11-21 所示。

```
description: Fast, reliable, scalable, and easy to use open-source relational database
  system.
engine: gotpl
home: https://www.mysql.com/
icon: https://www.mysql.com/common/logos/logo-mysql-170x115.png
keywords:
- mysql
- database
- sql
maintainers:
- email: viglesias@google.com
  name: Vic Iglesias
name: mysql
sources:
- https://github.com/kubernetes/charts
- https://github.com/docker-library/mysql
version: 0.3.0
```

图 11-21

name 和 version 是必填项,其他都是可选项。

(2) README.md

Markdown 格式的 README 文件,相当于 chart 的使用文档,此文件为可选,如图 11-22 所示。

```
# MySQL

[MySQL](https://MySQL.org) is one of the most popular database servers in the world. Notable users includ

## Introduction

This chart bootstraps a single node MySQL deployment on a [Kubernetes](http://kubernetes.io) cluster usin

## Prerequisites

- Kubernetes 1.6+ with Beta APIs enabled
- PV provisioner support in the underlying infrastructure

## Installing the Chart

To install the chart with the release name `my-release`:

```bash
$ helm install --name my-release stable/mysql
```

The command deploys MySQL on the Kubernetes cluster in the default configuration. The [configuration](#co

By default a random password will be generated for the root user. If you'd like to set your own password
in the values.yaml.

You can retrieve your root password by running the following command. Make sure to replace [YOUR_RELEASE_

    printf $(printf '\%o' `kubectl get secret [YOUR_RELEASE_NAME]-mysql -o jsonpath="{.data.mysql-root-pa

> **Tip**: List all releases using `helm list`

Uninstalling the Chart

To uninstall/delete the `my-release` deployment:

```bash
$ helm delete my-release
```

The command removes all the Kubernetes components associated with the chart and deletes the release.

Configuration

The following tables lists the configurable parameters of the MySQL chart and their default values.

Parameter	Description	Default
`imageTag`	`mysql` image tag.	Most recent release
`imagePullPolicy`	Image pull policy	`IfNotPresent`
`mysqlRootPassword`	Password for the `root` user.	`nil`
`mysqlUser`	Username of new user to create.	`nil`
`mysqlPassword`	Password for the new user.	`nil`
`mysqlDatabase`	Name for new database to create.	`nil`
```

图 11-22

（3）LICENSE

文本文件，描述 chart 的许可信息，此文件为可选。

（4）requirements.yaml

chart 可能依赖其他的 chart，这些依赖关系可通过 requirements.yaml 指定，如图 11-23 所示。

```
dependencies:
  - name: rabbitmq
    version: 1.2.3
    repository: http://example.com/charts
  - name: memcached
    version: 3.2.1
    repository: http://another.example.com/charts
```

图 11-23

在安装过程中，依赖的 chart 也会被一起安装。

（5）values.yaml

chart 支持在安装时根据参数进行定制化配置，而 values.yaml 则提供了这些配置参数的默认值，如图 11-24 所示。

```
## mysql image version
image: "mysql"
imageTag: "5.7.14"

## Specify password for root user
##
## Default: random 10 character string
# mysqlRootPassword: testing

## Create a database user
##
# mysqlUser:
# mysqlPassword:

imagePullPolicy: IfNotPresent

## Persist data to a persitent volume
persistence:
  enabled: true
  ## If defined, storageClassName: <storageClass>
  ## If set to "-", storageClassName: "", which disables dynamic provisioning
  ##
  # storageClass: "-"
  accessMode: ReadWriteOnce
  size: 8Gi

## Configure resource requests and limits
##
resources:
  requests:
    memory: 256Mi
    cpu: 100m

# Custom mysql configuration files used to override default mysql settings
configurationFiles:
#  mysql.cnf: |-
#    [mysqld]
#    skip-name-resolve
```

图 11-24

（6）templates 目录

各类 Kubernetes 资源的配置模板都放置在这里。Helm 会将 values.yaml 中的参数值注入模板中，生成标准的 YAML 配置文件。

模板是 chart 最重要的部分，也是 Helm 最强大的地方。模板增加了应用部署的灵活性，能够适用不同的环境，我们后面会详细讨论。

（7）templates/NOTES.txt

chart 的简易使用文档，chart 安装成功后会显示此文档内容，如图 11-25 所示。

```
MySQL can be accessed via port 3306 on the following DNS name from within your cluster:
{{ template "mysql.fullname" . }}.{{ .Release.Namespace }}.svc.cluster.local

To get your root password run:

    kubectl get secret --namespace {{ .Release.Namespace }} {{ template "mysql.fullname" . }}

To connect to your database:

1. Run an Ubuntu pod that you can use as a client:

    kubectl run -i --tty ubuntu --image=ubuntu:16.04 --restart=Never -- bash -il

2. Install the mysql client:

    $ apt-get update && apt-get install mysql-client -y

3. Connect using the mysql cli, then provide your password:
    $ mysql -h {{ template "mysql.fullname" . }} -p
```

图 11-25

与模板一样，可以在 NOTE.txt 中插入配置参数，Helm 会动态注入参数值。

11.5.2 chart 模板

Helm 通过模板创建 Kubernetes 能够理解的 YAML 格式的资源配置文件，我们将通过例子来学习如何使用模板。

以 templates/secrets.yaml 为例，如图 11-26 所示。

```
apiVersion: v1
kind: Secret
metadata:
  name: {{ template "mysql.fullname" . }}   ①
  labels:
    app: {{ template "mysql.fullname" . }}
    chart: "{{ .Chart.Name }}-{{ .Chart.Version }}"
    release: "{{ .Release.Name }}"              ②
    heritage: "{{ .Release.Service }}"
type: Opaque
data:
  {{ if .Values.mysqlRootPassword }}  ③
  mysql-root-password:  {{ .Values.mysqlRootPassword | b64enc | quote }}
  {{ else }}
  mysql-root-password: {{ randAlphaNum 10 | b64enc | quote }}
  {{ end }}
  {{ if .Values.mysqlPassword }}
  mysql-password:  {{ .Values.mysqlPassword | b64enc | quote }}
  {{ else }}
  mysql-password: {{ randAlphaNum 10 | b64enc | quote }}
  {{ end }}
```

图 11-26

从结构上看，文件的内容非常像 Secret 配置，只是大部分属性值变成了{{ xxx }}。这些{{ xxx }} 实际上是模板的语法。Helm 采用了 Go 语言的模板来编写 chart。Go 模板非常强大，支持变量、对象、函数、流控制等功能。下面我们通过解析 templates/secrets.yaml 快速学习模板。

① `{{ template "mysql.fullname" . }}` 定义 Secret 的 name。关键字 template 的作用是引用一个子模板 mysql.fullname。这个子模板是在 templates/_helpers.tpl 文件中定义的，如图 11-27 所示。

```
{{/* vim: set filetype=mustache: */}}
{{/*
Expand the name of the chart.
*/}}
{{- define "mysql.name" -}}
{{- default .Chart.Name .Values.nameOverride | trunc 63 | trimSuffix "-" -}}
{{- end -}}

{{/*
Create a default fully qualified app name.
We truncate at 63 chars because some Kubernetes name fields are limited to this
*/}}
{{- define "mysql.fullname" -}}
{{- $name := default .Chart.Name .Values.nameOverride -}}
{{- printf "%s-%s" .Release.Name $name | trunc 63 | trimSuffix "-" -}}
{{- end -}}
```

图 11-27

这个定义还是很复杂的，因为它用到了模板语言中的对象、函数、流控制等概念。现在看不懂没关系，这里我们学习的重点是：如果存在一些信息多个模板都会用到，则可在 templates/_helpers.tpl 中将其定义为子模板，然后通过 templates 函数引用。

这里 mysql.fullname 是由 release 与 chart 二者名字拼接组成的。

根据 chart 的最佳实践，所有资源的名称都应该保持一致。对于我们这个 chart，无论 Secret 还是 Deployment、PersistentVolumeClaim、Service，它们的名字都是子模板 mysql.fullname 的值。

② Chart 和 Release 是 Helm 预定义的对象，每个对象都有自己的属性，可以在模板中使用。如果使用下面的命令安装 chart：

```
helm install stable/mysql -n my
```

那么：

- `{{ .Chart.Name }}` 的值为 mysql。
- `{{ .Chart.Version }}` 的值为 0.3.0。
- `{{ .Release.Name }}` 的值为 my。
- `{{ .Release.Service }}` 始终取值为 Tiller。
- `{{ template "mysql.fullname" . }}` 计算结果为 my-mysql。

③ 这里指定 mysql-root-password 的值，不过使用了 if-else 的流控制，其逻辑为：如

果 .Values.mysqlRootPassword 有值，就对其进行 base64 编码，否则随机生成一个 10 位的字符串并编码。

Values 也是预定义的对象，代表的是 values.yaml 文件。.Values.mysqlRootPassword 则是 values.yaml 中定义的 mysqlRootPassword 参数，如图 11-28 所示。

图 11-28

因为 mysqlRootPassword 被注释掉了，没有赋值，所以逻辑判断会走 else，即随机生成密码。

randAlphaNum、b64enc、quote 都是 Go 模板语言支持的函数，函数之间可以通过管道 | 连接。{{ randAlphaNum 10 | b64enc | quote }} 的作用是首先随机产生一个长度为 10 的字符串，然后将其 base64 编码，最后两边加上双引号。

templates/secrets.yaml 这个例子展示了 chart 模板主要的功能，我们最大的收获应该是：模板将 chart 参数化了，通过 values.yaml 可以灵活定制应用。

无论多复杂的应用，用户都可以用 Go 模板语言编写出 chart。无非是使用到更多的函数、对象和流控制。对于初学者，建议尽量参考官方的 chart。根据二八定律，这些 chart 已经覆盖了绝大部分情况，而且采用了最佳实践。如何遇到不懂的函数、对象和其他语法，可参考官网文档 https://docs.helm.sh。

11.5.3 再次实践 MySQL chart

学习了 chart 结构和模板的知识后，现在重新实践一次 MySQL chart，相信会有更多收获。

1. chart 安装前的准备

作为准备工作，安装之前需要先清楚 chart 的使用方法。这些信息通常记录在 values.yaml 和 README.md 中。除了下载源文件查看，执行 helm inspect values 可能是更方便的方法，如图 11-29 所示。

```
ubuntu@k8s-master:~$
ubuntu@k8s-master:~$ helm inspect values stable/mysql
## mysql image version
## ref: https://hub.docker.com/r/library/mysql/tags/
##
image: "mysql"
imageTag: "5.7.14"

## Specify password for root user
##
## Default: random 10 character string
# mysqlRootPassword: testing

## Create a database user
##
# mysqlUser:
# mysqlPassword:

## Allow unauthenticated access, uncomment to enable
##
# mysqlAllowEmptyPassword: true
```

图 11-29

输出的实际上是 values.yaml 的内容。阅读注释就可以知道 MySQL chart 支持哪些参数，安装之前需要做哪些准备。其中有一部分是关于存储的，如图 11-30 所示。

```
persistence:
  enabled: true
  ## database data Persistent Volume Storage Class
  ## If defined, storageClassName: <storageClass>
  ## If set to "-", storageClassName: "", which disables dynamic provisioning
  ## If undefined (the default) or set to null, no storageClassName spec is
  ##   set, choosing the default provisioner. (gp2 on AWS, standard on
  ##   GKE, AWS & OpenStack)
  ##
  # storageClass: "-"
  accessMode: ReadWriteOnce
  size: 8Gi
```

图 11-30

chart 定义了一个 PersistentVolumeClaim，申请 8GB 的 PersistentVolume。由于我们的实验环境不支持动态供给，因此要预先创建好相应的 PV，其配置文件 mysql-pv.yml 的内容如图 11-31 所示。

```
apiVersion: v1
kind: PersistentVolume
metadata:
  name: mysql-pv
spec:
  accessModes:
    - ReadWriteOnce
  capacity:
    storage: 8Gi
  persistentVolumeReclaimPolicy: Retain
  # storageClassName: nfs
  nfs:
    path: /nfsdata/mysql-pv
    server: 192.168.56.105
```

图 11-31

创建 PV mysql-pv，如图 11-32 所示。

```
ubuntu@k8s-master:~$
ubuntu@k8s-master:~$ kubectl apply -f mysql-pv.yml
persistentvolume "mysql-pv" created
ubuntu@k8s-master:~$
ubuntu@k8s-master:~$ kubectl get pv
NAME       CAPACITY   ACCESSMODES   RECLAIMPOLICY   STATUS      CLAIM   STORAGECLASS
mysql-pv   8Gi        RWO           Retain          Available
ubuntu@k8s-master:~$
```

图 11-32

接下来就可以安装 chart 了。

2. 定制化安装 chart

除了接受 values.yaml 的默认值，我们还可以定制化 chart，比如设置 mysqlRootPassword。Helm 有两种方式传递配置参数：

（1）指定自己的 values 文件。通常的做法是首先通过 helm inspect values mysql > myvalues.yaml 生成 values 文件，然后设置 mysqlRootPassword，最后执行 helm install --values=myvalues.yaml mysql。

（2）通过 --set 直接传入参数值，如图 11-33 所示。

```
ubuntu@k8s-master:~$
ubuntu@k8s-master:~$ helm install stable/mysql --set mysqlRootPassword=abc123 -n my
NAME:   my
LAST DEPLOYED: Thu Oct 19 14:24:56 2017
NAMESPACE: default
STATUS: DEPLOYED

RESOURCES:
==> v1/Service
NAME      CLUSTER-IP      EXTERNAL-IP   PORT(S)    AGE
my-mysql  10.104.59.139   <none>        3306/TCP   0s

==> v1beta1/Deployment
NAME      DESIRED   CURRENT   UP-TO-DATE   AVAILABLE   AGE
my-mysql  1         1         1            0           0s

==> v1/Secret
NAME      TYPE     DATA   AGE
my-mysql  Opaque   2      0s

==> v1/PersistentVolumeClaim
NAME      STATUS    VOLUME   CAPACITY   ACCESSMODES   STORAGECLASS   AGE
my-mysql  Pending                                                    0s
```

图 11-33

mysqlRootPassword 设置为 abc123。另外，-n 设置 release 为 my，各类资源的名称即为 my-mysql。

通过 helm list 和 helm status 可以查看 chart 的最新状态，如图 11-34 所示。

图 11-34

PVC 已经 Bound，Deployment 也变为 AVAILABLE。

11.5.4　升级和回滚 release

release 发布后可以执行 helm upgrade 对其进行升级，通过 --values 或 --set 应用新的配置。比如将当前的 MySQL 版本升级到 5.7.15，如图 11-35 所示。

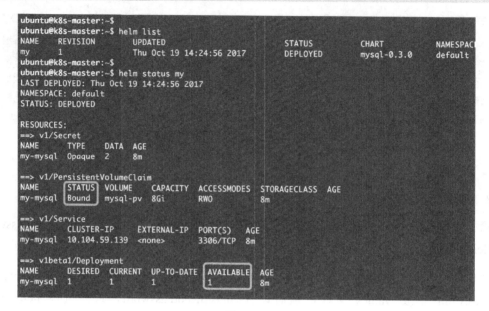

图 11-35

等待一些时间，升级成功，如图 11-36 所示。

图 11-36

helm history 可以查看 release 所有的版本。通过 helm rollback 可以回滚到任何版本，如图 11-37 所示。

图 11-37

回滚成功，MySQL 恢复到 5.7.14，如图 11-38 所示。

图 11-38

11.5.5 开发自己的 chart

Kubernetes 给我们提供了大量官方 chart，不过要部署微服务应用，还是需要开发自己的 chart，下面就来实践这个主题。

1. 创建 chart

执行 helm create mychart 命令，创建 chart mychart，如图 11-39 所示。

图 11-39

Helm 会帮我们创建目录 mychart，并生成各类 chart 文件。这样我们就可以在此基础上开发自己的 chart 了。

新建的 chart 默认包含一个 nginx 应用示例，values.yaml 内容如图 11-40 所示。

```
# Default values for mychart.
# This is a YAML-formatted file.
# Declare variables to be passed into your templates.
replicaCount 1
image:
  repository: nginx
  tag: stable
  pullPolicy: IfNotPresent
service:
  name: nginx
  type: ClusterIP
  externalPort: 80
  internalPort: 80
ingress:
  enabled: false
  # Used to create an Ingress record.
  hosts:
    - chart-example.local
```

图 11-40

开发时建议大家参考官方 chart 中的模板 values.yaml、Chart.yaml，里面包含了大量最佳实践和最常用的函数、流控制，这里就不一一展开了。

2. 调试 chart

只要是程序就会有 bug，chart 也不例外。Helm 提供了 debug 的工具：helm lint 和 helm install --dry-run --debug。

helm lint 会检测 chart 的语法，报告错误以及给出建议。

比如我们故意在 values.yaml 的第 8 行漏掉了行尾的那个冒号，如图 11-41 所示。

```
 1 # Default values for mychart.
 2 # This is a YAML-formatted file.
 3 # Declare variables to be passed into your templates.
 4 replicaCount: 1
 5 image:
 6   repository: nginx
 7   tag: stable
 8   pullPolicy IfNotPresent
 9 service:
10   name: nginx
11   type: ClusterIP
12   externalPort: 80
13   internalPort: 80
```

图 11-41

helm lint mychart 会指出这个语法错误，如图 11-42 所示。

```
ubuntu@k8s-master:~$
ubuntu@k8s-master:~$ helm lint mychart
==> Linting mychart
[INFO] Chart.yaml: icon is recommended
[ERROR] values.yaml: unable to parse YAML
        error converting YAML to JSON: yaml: line 8: could not find expected ':'

Error: 1 chart(s) linted, 1 chart(s) failed
ubuntu@k8s-master:~$
```

图 11-42

mychart 目录被作为参数传递给 helm lint。错误修复后则能通过检测，如图 11-43 所示。

```
ubuntu@k8s-master:~$
ubuntu@k8s-master:~$ helm lint mychart
==> Linting mychart
[INFO] Chart.yaml: icon is recommended

1 chart(s) linted, no failures
ubuntu@k8s-master:~$
```

图 11-43

helm install --dry-run --debug 会模拟安装 chart，并输出每个模板生成的 YAML 内容，如图 11-44、图 11-45 所示。

```
ubuntu@k8s-master:~$
ubuntu@k8s-master:~$ helm install --dry-run mychart --debug
[debug] Created tunnel using local port: '46295'

[debug] SERVER: "localhost:46295"

[debug] Original chart version: ""
[debug] CHART PATH: /home/ubuntu/mychart

NAME:   solitary-kudu
REVISION: 1
RELEASED: Fri Oct 20 09:59:51 2017
CHART: mychart-0.1.0
USER-SUPPLIED VALUES:
{}

COMPUTED VALUES:
image:
  pullPolicy: IfNotPresent
  repository: nginx
  tag: stable
ingress:
  annotations: null
  enabled: false
  hosts:
  - chart-example.local
  tls: null
replicaCount: 1
resources: {}
service:
  externalPort: 80
  internalPort: 80
  name: nginx
  type: ClusterIP

HOOKS:
MANIFEST:
```

图 11-44

```yaml
---
# Source: mychart/templates/service.yaml
apiVersion: v1
kind: Service
metadata:
  name: solitary-kudu-mychart
  labels:
    app: mychart
    chart: mychart-0.1.0
    release: solitary-kudu
    heritage: Tiller
spec:
  type: ClusterIP
  ports:
    - port: 80
      targetPort: 80
      protocol: TCP
      name: nginx
  selector:
    app: mychart
    release: solitary-kudu
---
# Source: mychart/templates/deployment.yaml
apiVersion: extensions/v1beta1
kind: Deployment
metadata:
  name: solitary-kudu-mychart
  labels:
    app: mychart
    chart: mychart-0.1.0
    release: solitary-kudu
    heritage: Tiller
spec:
  replicas: 1
  template:
    metadata:
      labels:
        app: mychart
        release: solitary-kudu
    spec:
      containers:
        - name: mychart
          image: "nginx:stable"
          imagePullPolicy: IfNotPresent
          ports:
            - containerPort: 80
          livenessProbe:
            httpGet:
              path: /
              port: 80
          readinessProbe:
            httpGet:
              path: /
              port: 80
          resources:
            {}
```

图 11-45

我们可以检测这些输出，判断是否与预期相符。

同样，mychart 目录作为参数传递给 helm install --dry-run --debug。

3. 安装 chart

当我们准备就绪，就可以安装 chart 了。Helm 支持四种安装方法：

（1）安装仓库中的 chart，例如 helm install stable/nginx。
（2）通过 tar 包安装，例如 helm install ./nginx-1.2.3.tgz。
（3）通过 chart 本地目录安装，例如 helm install ./nginx。
（4）通过 URL 安装，例如 helm install https://example.com/charts/nginx-1.2.3.tgz。

这里我们使用本地目录安装，如图 11-46 所示。

```
ubuntu@k8s-master:~$
ubuntu@k8s-master:~$ helm install mychart
NAME:    opinionated-nightingale
LAST DEPLOYED: Fri Oct 20 10:14:06 2017
NAMESPACE: default
STATUS: DEPLOYED

RESOURCES:
==> v1/Service
NAME                              CLUSTER-IP       EXTERNAL-IP   PORT(S)   AGE
opinionated-nightingale-mychart   10.100.28.221    <none>        80/TCP    0s

==> v1beta1/Deployment
NAME                              DESIRED   CURRENT   UP-TO-DATE   AVAILABLE   AGE
opinionated-nightingale-mychart   1         1         1            0           0s


NOTES:
1. Get the application URL by running these commands:
   export POD_NAME=$(kubectl get pods --namespace default -l "app=mychart,release=opi
   echo "Visit http://127.0.0.1:8080 to use your application"
   kubectl port-forward $POD_NAME 8080:80

ubuntu@k8s-master:~$
```

图 11-46

当 chart 部署到 Kubernetes 集群后，便可以对其进行更为全面的测试了。

4. 将 chart 添加到仓库

chart 通过测试后可以添加到仓库中，团队其他成员就能够使用了。任何 HTTP Server 都可以用作 chart 仓库。下面演示在 k8s-node1 192.168.56.106 上搭建仓库。

（1）在 k8s-node1 上启动一个 httpd 容器，如图 11-47 所示。

```
root@k8s-node1:~#
root@k8s-node1:~# mkdir /var/www
root@k8s-node1:~#
root@k8s-node1:~# docker run -d -p 8080:80 -v /var/www/:/usr/local/apache2/htdocs/ httpd
f3d5f0779a04a3074bd332764263e0283d8548e8f9b2b96dc744e098b45ce075
root@k8s-node1:~#
```

图 11-47

（2）通过 helm package 将 mychart 打包，如图 11-48 所示。

```
ubuntu@k8s-master:~$
ubuntu@k8s-master:~$ helm package mychart
Successfully packaged chart and saved it to: /home/ubuntu/mychart-0.1.0.tgz
ubuntu@k8s-master:~$
```

图 11-48

（3）执行 helm repo index 生成仓库的 index 文件，如图 11-49 所示。

```
ubuntu@k8s-master:~$
ubuntu@k8s-master:~$ mkdir myrepo
ubuntu@k8s-master:~$ mv mychart-0.1.0.tgz myrepo/
ubuntu@k8s-master:~$ helm repo index myrepo/ --url http://192.168.56.106:8080/charts
ubuntu@k8s-master:~$ ls myrepo/
index.yaml  mychart-0.1.0.tgz
ubuntu@k8s-master:~$
```

图 11-49

Helm 会扫描 myrepo 目录中的所有 tgz 包并生成 index.yaml。--url 指定的是新仓库的访问路径。新生成的 index.yaml 记录了当前仓库中所有 chart 的信息，如图 11-50 所示。

```
apiVersion: v1
entries:
  mychart:
  - apiVersion: v1
    created: 2017-10-21T16:39:39.184818935+08:00
    description: A Helm chart for Kubernetes
    digest: ae8d7138002d432014dc8638ec37202823e9207445caf08a660d154b26e936ea
    name: mychart
    urls:
    - http://192.168.56.106:8080/charts/mychart-0.1.0.tgz
    version: 0.1.0
generated: 2017-10-21T16:39:39.184265707+08:00
```

图 11-50

当前只有 mychart 这一个 chart。

（4）将 mychart-0.1.0.tgz 和 index.yaml 上传到 k8s-node1 的 /var/www/charts 目录，如图 11-51 所示。

```
root@k8s-node1:~#
root@k8s-node1:~# ls /var/www/charts/
index.yaml  mychart-0.1.0.tgz
root@k8s-node1:~#
```

图 11-51

（5）通过 helm repo add 将新仓库添加到 Helm，如图 11-52 所示。

```
ubuntu@k8s-master:~$
ubuntu@k8s-master:~$ helm repo add newrepo http://192.168.56.106:8080/charts
"newrepo" has been added to your repositories
ubuntu@k8s-master:~$
ubuntu@k8s-master:~$ helm repo list
NAME       URL
stable     https://kubernetes-charts.storage.googleapis.com
local      http://127.0.0.1:8879/charts
newrepo    http://192.168.56.106:8080/charts
ubuntu@k8s-master:~$
```

图 11-52

仓库命名为 newrepo，Helm 会从仓库下载 index.yaml。

（6）现在已经可以 repo search 到 mychart 了，如图 11-53 所示。

```
ubuntu@k8s-master:~$
ubuntu@k8s-master:~$ helm search mychart
NAME              VERSION  DESCRIPTION
local/mychart     0.1.0    A Helm chart for Kubernetes
newrepo/mychart   0.1.0    A Helm chart for Kubernetes
ubuntu@k8s-master:~$
```

图 11-53

除了 newrepo/mychart，这里还有一个 local/mychart。这是因为在执行第 2 步打包操作的同时，mychart 也被同步到了 local 的仓库。

（7）已经可以直接从新仓库安装 mychart 了，如图 11-54 所示。

```
ubuntu@k8s-master:~$
ubuntu@k8s-master:~$ helm install newrepo/mychart
NAME:   guilded-jellyfish
LAST DEPLOYED: Sat Oct 21 16:49:31 2017
NAMESPACE: default
STATUS: DEPLOYED

RESOURCES:
==> v1/Service
NAME                        CLUSTER-IP    EXTERNAL-IP  PORT(S)  AGE
guilded-jellyfish-mychart   10.109.0.87   <none>       80/TCP   0s

==> v1beta1/Deployment
NAME                        DESIRED  CURRENT  UP-TO-DATE  AVAILABLE  AGE
guilded-jellyfish-mychart   1        1        1           0          0s


NOTES:
1. Get the application URL by running these commands:
   export POD_NAME=$(kubectl get pods --namespace default -l "app=mychart,r
   echo "Visit http://127.0.0.1:8080 to use your application"
   kubectl port-forward $POD_NAME 8080:80
```

图 11-54

（8）若以后仓库添加了新的 chart，则需要用 helm repo update 更新本地的 index，如图 11-55 所示。

```
ubuntu@k8s-master:~$
ubuntu@k8s-master:~$ helm repo update
Hang tight while we grab the latest from your chart repositories...
...Skip local chart repository
...Successfully got an update from the "newrepo" chart repository
...Successfully got an update from the "stable" chart repository
Update Complete. ⎈ Happy Helming!⎈
ubuntu@k8s-master:~$
```

图 11-55

这个操作相当于 Ubutun 的 apt-get update。

11.6 小结

本章我们学习了 Kubernetes 包管理器 Helm。

Helm 让我们能够像 apt 管理 deb 包那样安装、部署、升级和删除容器化应用。

Helm 由客户端和 Tiller 服务器组成。客户端负责管理 chart，服务器负责管理 release。

chart 是 Helm 的应用打包格式，它由一组文件和目录构成。其中最重要的是模板，模板中定义了 Kubernetes 各类资源的配置信息，Helm 在部署时通过 values.yaml 实例化模板。

Helm 允许用户开发自己的 chart，并为用户提供了调试工具。用户可以搭建自己的 chart 仓库，在团队中共享 chart。

Helm 帮助用户在 Kubernetes 上高效地运行和管理微服务架构应用，Helm 非常重要。

第 12 章
网　络

本章我们讨论 Kubernetes 网络这个重要主题。

Kubernetes 作为编排引擎管理着分布在不同节点上的容器和 Pod。Pod、Service、外部组件之间需要一种可靠的方式找到彼此并进行通信，Kubernetes 网络则负责提供这个保障。本章包括如下内容：

（1）Kubernetes 网络模型。
（2）各种网络方案。
（3）Network Policy。

12.1　Kubernetes 网络模型

Kubernetes 采用的是基于扁平地址空间的网络模型，集群中的每个 Pod 都有自己的 IP 地址，Pod 之间不需要配置 NAT 就能直接通信。另外，同一个 Pod 中的容器共享 Pod 的 IP，能够通过 localhost 通信。

这种网络模型对应用开发者和管理员相当友好，应用可以非常方便地从传统网络迁移到 Kubernetes。每个 Pod 可被看作是一个个独立的系统，而 Pod 中的容器则可被看作同一系统中的不同进程。

下面讨论在这个网络模型下集群中的各种实体如何通信。知识点前面都已经涉及，这里可当作复习和总结。

1. Pod 内容器之间的通信

当 Pod 被调度到某个节点，Pod 中的所有容器都在这个节点上运行，这些容器共享相同的本地文件系统、IPC 和网络命名空间。

不同 Pod 之间不存在端口冲突的问题，因为每个 Pod 都有自己的 IP 地址。当某个容器使用 localhost 时，意味着使用的是容器所属 Pod 的地址空间。

比如 Pod A 有两个容器 container-A1 和 container-A2，container-A1 在端口 1234 上监听，当 container-A2 连接到 localhost:1234 时，实际上就是在访问 container-A1。这不会与同一个节点上的 Pod B 冲突，即使 Pod B 中的容器 container-B1 也在监听 1234 端口。

2. Pod 之间的通信

Pod 的 IP 是集群可见的，即集群中的任何其他 Pod 和节点都可以通过 IP 直接与 Pod 通信，这种通信不需要借助任何网络地址转换、隧道或代理技术。Pod 内部和外部使用的是同一个 IP，这也意味着标准的命名服务和发现机制，比如 DNS 可以直接使用。

3. Pod 与 Service 的通信

Pod 间可以直接通过 IP 地址通信，但前提是 Pod 知道对方的 IP。在 Kubernetes 集群中，Pod 可能会频繁地销毁和创建，也就是说 Pod 的 IP 不是固定的。为了解决这个问题，Service 提供了访问 Pod 的抽象层。无论后端的 Pod 如何变化，Service 都作为稳定的前端对外提供服务。同时，Service 还提供了高可用和负载均衡功能，Service 负责将请求转发给正确的 Pod。

4. 外部访问

无论是 Pod 的 IP 还是 Service 的 Cluster IP，它们只能在 Kubernetes 集群中可见，对集群之外的世界，这些 IP 都是私有的。

Kubernetes 提供了两种方式让外界能够与 Pod 通信：

- NodePort。Service 通过 Cluster 节点的静态端口对外提供服务。外部可以通过 <NodeIP>:<NodePort> 访问 Service。
- LoadBalancer。Service 利用 cloud provider 提供的 load balancer 对外提供服务，cloud provider 负责将 load balancer 的流量导向 Service。目前支持的 cloud provider 有 GCP、AWS、Azur 等。

12.2 各种网络方案

网络模型有了，如何实现呢？

为了保证网络方案的标准化、扩展性和灵活性，Kubernetes 采用了 Container Networking Interface（CNI）规范。

CNI 是由 CoreOS 提出的容器网络规范，使用了插件（Plugin）模型创建容器的网络栈，如图 12-1 所示。

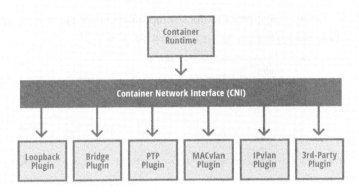

图 12-1

CNI 的优点是支持多种容器 runtime，不仅仅是 Docker。CNI 的插件模型支持不同组织和公司开发的第三方插件，这对运维人员来说很有吸引力，可以灵活选择适合的网络方案。

目前已有多种支持 Kubernetes 的网络方案，比如 Flannel、Calico、Canal、Weave Net 等。因为它们都实现了 CNI 规范，用户无论选择哪种方案，得到的网络模型都一样，即每个 Pod 都有独立的 IP，可以直接通信。区别在于不同方案的底层实现不同，有的采用基于 VxLAN 的 Overlay 实现，有的则是 Underlay，性能上有区别。再有就是是否支持 Network Policy。

12.3 Network Policy

Network Policy 是 Kubernetes 的一种资源。Network Policy 通过 Label 选择 Pod，并指定其他 Pod 或外界如何与这些 Pod 通信。

默认情况下，所有 Pod 是非隔离的，即任何来源的网络流量都能够访问 Pod，没有任何限制。当为 Pod 定义了 Network Policy 时，只有 Policy 允许的流量才能访问 Pod。

不过，不是所有的 Kubernetes 网络方案都支持 Network Policy。比如 Flannel 就不支持，Calico 是支持的。我们接下来将用 Canal 来演示 Network Policy。Canal 这个开源项目很有意思，它用 Flannel 实现 Kubernetes 集群网络，同时又用 Calico 实现 Network Policy。

12.3.1 部署 Canal

部署 Canal 与部署其他 Kubernetes 网络方案非常类似，都是在执行了 kubeadm init 初始化 Kubernetes 集群之后通过 kubectl apply 安装相应的网络方案。也就是说，没有太好的办法直接切换使用不同的网络方案，基本上只能重新创建集群。

要销毁当前集群，最简单的方法是在每个节点上执行 kubeadm reset，然后就可以按照 3.3.1 小节"初始化 Master"中的方法初始化集群了。

```
kubeadm init --apiserver-advertise-address 192.168.56.105
```

```
--pod-network-cidr=10.244.0.0/16
```

然后按照文档 https://kubernetes.io/docs/setup/independent/create-cluster-kubeadm/ 安装 Canal。文档列出了各种网络方案的安装方法，如图 12-2 所示。

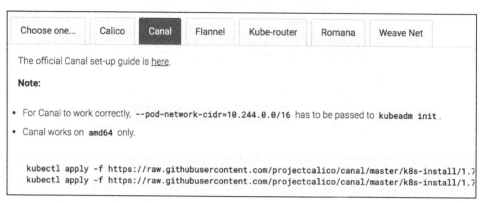

图 12-2

执行如下命令部署 Canal：

```
kubectl apply -f https://raw.githubusercontent.com/projectcalico/canal/master/k8s-install/1.7/rbac.yaml
kubectl apply -f https://raw.githubusercontent.com/projectcalico/canal/master/k8s-install/1.7/canal.yaml
```

部署成功后，可以查看到 Canal 相关组件，如图 12-3 所示。

```
ubuntu@k8s-master:~$
ubuntu@k8s-master:~$ kubectl get --namespace=kube-system daemonset canal
NAME    DESIRED   CURRENT   READY   UP-TO-DATE   AVAILABLE   NODE SELECTOR   AGE
canal   3         3         3       3            3           <none>          5d
ubuntu@k8s-master:~$
ubuntu@k8s-master:~$ kubectl get --namespace=kube-system pod -o wide|grep canal
canal-fkrl8                         3/3       Running   0          5d        192.168.56.107   k8s-node2
canal-pqvq2                         3/3       Running   0          5d        192.168.56.105   k8s-master
canal-wtlhk                         3/3       Running   0          5d        192.168.56.106   k8s-node1
ubuntu@k8s-master:~$
```

图 12-3

Canal 作为 DaemonSet 部署到每个节点，属于 kube-system 这个 namespace。

12.3.2 实践 Network Policy

为了演示 Network Policy，我们先部署一个 httpd 应用，其配置文件 httpd.yaml 如图 12-4 所示。

```yaml
apiVersion: apps/v1beta1
kind: Deployment
metadata:
  name: httpd
spec:
  replicas: 3
  template:
    metadata:
      labels:
        run: httpd
    spec:
      containers:
      - name: httpd
        image: httpd:latest
        imagePullPolicy: IfNotPresent
        ports:
        - containerPort: 80
---
apiVersion: v1
kind: Service
metadata:
  name: httpd-svc
spec:
  type: NodePort
  selector:
    run: httpd
  ports:
  - protocol: TCP
    nodePort: 30000
    port: 8080
    targetPort: 80
```

图 12-4

httpd 有三个副本，通过 NodePort 类型的 Service 对外提供服务。部署应用，如图 12-5 所示。

```
ubuntu@k8s-master:~$
ubuntu@k8s-master:~$ kubectl apply -f httpd.yml
deployment "httpd" created
service "httpd-svc" created
ubuntu@k8s-master:~$
ubuntu@k8s-master:~$ kubectl get pod -o wide
NAME                   READY   STATUS    RESTARTS   AGE   IP           NODE
httpd-b5c6f48-gs7p2    1/1     Running   0          2m    10.244.1.7   k8s-node1
httpd-b5c6f48-p86tv    1/1     Running   0          2m    10.244.1.8   k8s-node1
httpd-b5c6f48-qxzxg    1/1     Running   0          2m    10.244.2.4   k8s-node2
ubuntu@k8s-master:~$
ubuntu@k8s-master:~$ kubectl get service httpd-svc
NAME        TYPE       CLUSTER-IP       EXTERNAL-IP   PORT(S)          AGE
httpd-svc   NodePort   10.104.77.239    <none>        8080:30000/TCP   2m
ubuntu@k8s-master:~$
```

图 12-5

当前没有定义任何 Network Policy，验证应用可以被访问，如图 12-6 所示。

（1）启动一个 busybox Pod，既可以访问 Service，也可以 Ping 到副本 Pod。

```
ubuntu@k8s-master:~$
ubuntu@k8s-master:~$ kubectl run busybox --rm -ti --image=busybox /bin/sh
If you don't see a command prompt, try pressing enter.
/ #
/ # wget httpd-svc:8080
Connecting to httpd-svc:8080 (10.104.77.239:8080)
index.html           100% |*********************************************
/ #
/ # ping 10.244.1.7
PING 10.244.1.7 (10.244.1.7): 56 data bytes
64 bytes from 10.244.1.7: seq=0 ttl=62 time=1.696 ms
64 bytes from 10.244.1.7: seq=1 ttl=62 time=0.558 ms
64 bytes from 10.244.1.7: seq=2 ttl=62 time=0.497 ms
```

图 12-6

（2）集群节点既可以访问 Service，也可以 Ping 到副本 Pod，如图 12-7 所示。

```
root@k8s-node2:~#
root@k8s-node2:~# curl 10.104.77.239:8080
<html><body><h1>It works!</h1></body></html>
root@k8s-node2:~#
root@k8s-node2:~# ping -c 3 10.244.1.7
PING 10.244.1.7 (10.244.1.7) 56(84) bytes of data.
64 bytes from 10.244.1.7: icmp_seq=1 ttl=63 time=0.487 ms
64 bytes from 10.244.1.7: icmp_seq=2 ttl=63 time=0.494 ms
64 bytes from 10.244.1.7: icmp_seq=3 ttl=63 time=0.495 ms
```

图 12-7

（3）集群外（192.168.56.1）可以访问 Service，如图 12-8 所示。

```
➜  ~
➜  ~ curl 192.168.56.106:30000
<html><body><h1>It works!</h1></body></html>
➜  ~
```

图 12-8

现在创建 Network Policy，如图 12-9 所示。

```
kind: NetworkPolicy
apiVersion: networking.k8s.io/v1
metadata:
  name: access-httpd
spec:
  podSelector:
    matchLabels:
      run: httpd      ①
  ingress:
  - from:
    - podSelector:
        matchLabels:
          access: "true"  ②
    ports:
    - protocol: TCP
      port: 80      ③
```

图 12-9

① 定义将此 Network Policy 中的访问规则应用于 label 为 run: httpd 的 Pod，即 httpd

应用的三个副本 Pod。

② ingress 中定义只有 label 为 access: "true" 的 Pod 才能访问应用。

③ 只能访问 80 端口。

通过 kubectl apply 创建 Network Policy，如图 12-10 所示。

```
ubuntu@k8s-master:~$
ubuntu@k8s-master:~$ kubectl apply -f policy.yaml
networkpolicy "access-httpd" created
ubuntu@k8s-master:~$
ubuntu@k8s-master:~$ kubectl get networkpolicy
NAME               POD-SELECTOR    AGE
access-httpd       run=httpd       32s
ubuntu@k8s-master:~$
```

图 12-10

验证 Network Policy 的有效性：

（1）busybox Pod 已经不能访问 Service，如图 12-11 所示。

```
ubuntu@k8s-master:~$
ubuntu@k8s-master:~$ kubectl run busybox --rm -ti --image=busybox /bin/sh
If you don't see a command prompt, try pressing enter.
/ #
/ # wget httpd-svc:8080 --timeout=5
Connecting to httpd-svc:8080 (10.104.77.239:8080)
wget: download timed out
/ #
```

图 12-11

如果 Pod 添加了 label access: "true" 就能访问到应用，但 Ping 已经被禁止，如图 12-12 所示。

```
ubuntu@k8s-master:~$
ubuntu@k8s-master:~$ kubectl run busybox --rm -ti --labels="access=true" --image=busybox /bin/sh
If you don't see a command prompt, try pressing enter.
/ #
/ # wget httpd-svc:8080
Connecting to httpd-svc:8080 (10.104.77.239:8080)
index.html           100% |*******************************************|    45  0:00:00
/ #
/ # ping -c 3 10.244.1.7
PING 10.244.1.7 (10.244.1.7): 56 data bytes

--- 10.244.1.7 ping statistics ---
3 packets transmitted, 0 packets received, 100% packet loss
/ #
```

图 12-12

（2）集群节点已经不能访问 Service，也 Ping 不到副本 Pod，如图 12-13 所示。

```
root@k8s-node2:~#
root@k8s-node2:~# curl 10.104.77.239:8080 --connect-timeout 5
curl: (28) Connection timed out after 5001 milliseconds
root@k8s-node2:~#
root@k8s-node2:~# ping -c 3 10.244.1.7
PING 10.244.1.7 (10.244.1.7) 56(84) bytes of data.

--- 10.244.1.7 ping statistics ---
3 packets transmitted, 0 received, 100% packet loss, time 1999ms
```

图 12-13

（3）集群外（192.168.56.1）已经不能访问 Service，如图 12-14 所示。

```
➜  ~
➜  ~ curl 192.168.56.106:30000 --connect-timeout 5
curl: (28) Connection timed out after 5003 milliseconds
➜  ~
```

图 12-14

如果希望让集群节点和集群外（192.168.56.1）也能够访问到应用，可以对 Network Policy 做如图 12-15 所示的修改。

```yaml
kind: NetworkPolicy
apiVersion: networking.k8s.io/v1
metadata:
  name: access-httpd
spec:
  podSelector:
    matchLabels:
      run: httpd
  ingress:
  - from:
    - podSelector:
        matchLabels:
          access: "true"
    - ipBlock:
        cidr: 192.168.56.0/24
    ports:
    - protocol: TCP
      port: 80
```

图 12-15

应用新的 Network Policy，如图 12-16 所示。

```
ubuntu@k8s-master:~$
ubuntu@k8s-master:~$ kubectl apply -f policy.yaml
networkpolicy "access-httpd" configured
ubuntu@k8s-master:~$
```

图 12-16

现在，集群节点和集群外（192.168.56.1）已经能够访问了，如图 12-17、图 12-18 所示。

```
root@k8s-node2:~#
root@k8s-node2:~# curl 10.104.77.239:8080
<html><body><h1>It works!</h1></body></html>
root@k8s-node2:~#
```

图 12-17

```
➜  ~
➜  ~ curl 192.168.56.106:30000
<html><body><h1>It works!</h1></body></html>
➜  ~
```

图 12-18

除了通过 ingress 限制进入的流量，也可以用 egress 限制外出的流量。大家可以参考官网相关文档和示例，这里就不赘述了。

12.4 小结

Kubernetes 采用的是扁平化的网络模型，每个 Pod 都有自己的 IP，并且可以直接通信。
CNI 规范使得 Kubernetes 可以灵活选择多种 Plugin 实现集群网络。
Network Policy 赋予了 Kubernetes 强大的网络访问控制机制。

第 13 章
Kubernetes Dashboard

前面章节 Kubernetes 所有的操作我们都是通过命令行工具 kubectl 完成的。为了提供更丰富的用户体验，Kubernetes 还开发了一个基于 Web 的 Dashboard，用户可以用 Kubernetes Dashboard 部署容器化的应用、监控应用的状态、执行故障排查任务以及管理 Kubernetes 的各种资源。

在 Kubernetes Dashboard 中可以查看集群中应用的运行状态，也能够创建和修改各种 Kubernetes 资源，比如 Deployment、Job、DaemonSet 等。用户可以 Scale Up/Down Deployment、执行 Rolling Update、重启某个 Pod 或者通过向导部署新的应用。Dashboard 能显示集群中各种资源的状态以及日志信息。

可以说，Kubernetes Dashboard 提供了 kubectl 的绝大部分功能，大家可以根据情况进行选择。

13.1 安装

Kubernetes 默认没有部署 Dashboard，可通过如下命令安装：

```
kubectl create -f
https://raw.githubusercontent.com/kubernetes/dashboard/master/src/deploy/recommended/kubernetes-dashboard.yaml
```

Dashboard 会在 kube-system namespace 中创建自己的 Deployment 和 Service，如图 13-1 所示。

```
ubuntu@k8s-master:~$ kubectl --namespace=kube-system get deployment kubernetes-dashboard
NAME                   DESIRED   CURRENT   UP-TO-DATE   AVAILABLE   AGE
kubernetes-dashboard   1         1         1            1           17h
ubuntu@k8s-master:~$
ubuntu@k8s-master:~$ kubectl --namespace=kube-system get service kubernetes-dashboard
NAME                   TYPE        CLUSTER-IP       EXTERNAL-IP   PORT(S)   AGE
kubernetes-dashboard   ClusterIP   10.110.199.111   <none>        443/TCP   17h
```

图 13-1

因为 Service 是 ClusterIP 类型，为了方便使用，我们可通过 kubectl --namespace=kube-system edit service kubernetes-dashboard 修改成 NodePort 类型，如图 13-2 所示。

```
spec:
  clusterIP: 10.110.199.111
  ports:
  - port: 443
    protocol: TCP
    targetPort: 8443
  selector:
    k8s-app: kubernetes-dashboard
  sessionAffinity: None
  type: NodePort
status:
  loadBalancer: {}
```

图 13-2

保存修改，此时已经为 Service 分配了端口 31614，如图 13-3 所示。

```
ubuntu@k8s-master:~$
ubuntu@k8s-master:~$ kubectl --namespace=kube-system get service kubernetes-dash
NAME                   TYPE       CLUSTER-IP       EXTERNAL-IP   PORT(S)
kubernetes-dashboard   NodePort   10.110.199.111   <none>        443:31614/TCP
ubuntu@k8s-master:~$
```

图 13-3

通过浏览器访问 Dashboard https://192.168.56.105:31614/，登录界面如图 13-4 所示。

Kubernetes Dashboard

Authentication method:

◉ Kubeconfig

○ Token

Kubeconfig YAML file *

SIGN IN   SKIP

图 13-4

13.2 配置登录权限

Dashboard 支持 Kubeconfig 和 Token 两种认证方式，为了简化配置，我们通过配置文件 dashboard-admin.yaml 为 Dashboard 默认用户赋予 admin 权限，如图 13-5 所示。

```yaml
apiVersion: rbac.authorization.k8s.io/v1beta1
kind: ClusterRoleBinding
metadata:
  name: kubernetes-dashboard
  labels:
    k8s-app: kubernetes-dashboard
roleRef:
  apiGroup: rbac.authorization.k8s.io
  kind: ClusterRole
  name: cluster-admin
subjects:
- kind: ServiceAccount
  name: kubernetes-dashboard
  namespace: kube-system
```

图 13-5

执行 kubectl apply 使之生效，如图 13-6 所示。

```
ubuntu@k8s-master:~$
ubuntu@k8s-master:~$ kubectl apply -f dashboard-admin.yaml
clusterrolebinding "kubernetes-dashboard" created
ubuntu@k8s-master:~$
```

图 13-6

现在直接单击登录页面中的 SKIP 就可以进入 Dashboard 了，如图 13-7 所示。

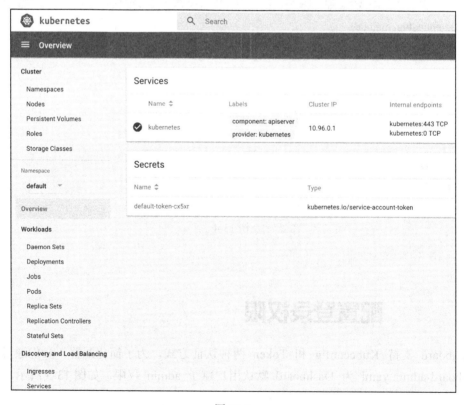

图 13-7

13.3 Dashboard 界面结构

Dashboard 的界面很简洁，分为三个大的区域。

（1）顶部操作区，如图 13-8 所示。在这里用户可以搜索集群中的资源、创建资源或退出。

图 13-8

（2）左边导航菜单。通过导航菜单可以查看和管理集群中的各种资源。菜单项按照资源的层级分为以下两类：

- Cluster 级别的资源，如图 13-9 所示。
- Namespace 级别的资源，如图 13-10 所示。

默认显示的是 default Namespace，可以进行切换，如图 13-11 所示。

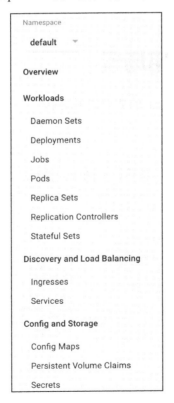

图 13-9　　　　图 13-10　　　　图 13-11

（3）中间主体区。

在导航菜单中单击了某类资源，中间主体区就会显示该资源所有实例，比如单击 Pods，如图 13-12 所示。

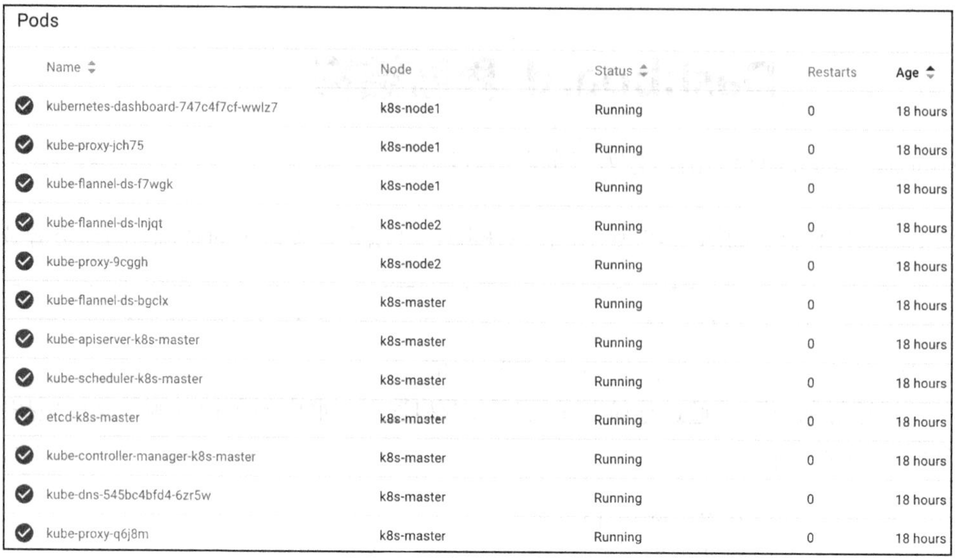

Pods				
Name	Node	Status	Restarts	Age
✓ kubernetes-dashboard-747c4f7cf-wwlz7	k8s-node1	Running	0	18 hours
✓ kube-proxy-jch75	k8s-node1	Running	0	18 hours
✓ kube-flannel-ds-f7wgk	k8s-node1	Running	0	18 hours
✓ kube-flannel-ds-lnjqt	k8s-node2	Running	0	18 hours
✓ kube-proxy-9cggh	k8s-node2	Running	0	18 hours
✓ kube-flannel-ds-bgclx	k8s-master	Running	0	18 hours
✓ kube-apiserver-k8s-master	k8s-master	Running	0	18 hours
✓ kube-scheduler-k8s-master	k8s-master	Running	0	18 hours
✓ etcd-k8s-master	k8s-master	Running	0	18 hours
✓ kube-controller-manager-k8s-master	k8s-master	Running	0	18 hours
✓ kube-dns-545bc4bfd4-6zr5w	k8s-master	Running	0	18 hours
✓ kube-proxy-q6j8m	k8s-master	Running	0	18 hours

图 13-12

13.4 典型使用场景

接下来我们介绍几个 Dashboard 的典型使用场景。

13.4.1 部署 Deployment

单击顶部操作区的 +CREATE 按钮，如图 13-13 所示。

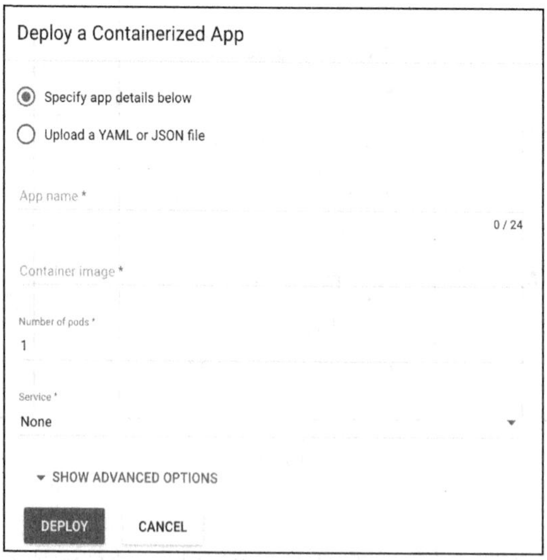

图 13-13

用户可以直接输入要部署应用的名字、镜像、副本数等信息,也可以上传 YAML 配置文件。如果是上传配置文件,则可以创建任意类型的资源,而不仅仅是 Deployment。

13.4.2 在线操作

对于每种资源,都可以单击按钮执行各种操作,如图 13-14 所示。

图 13-14

例如,单击 View/edit YAML,可直接修改资源的配置,保存后立即生效,其效果与 kubectl edit 一样,如图 13-15 所示。

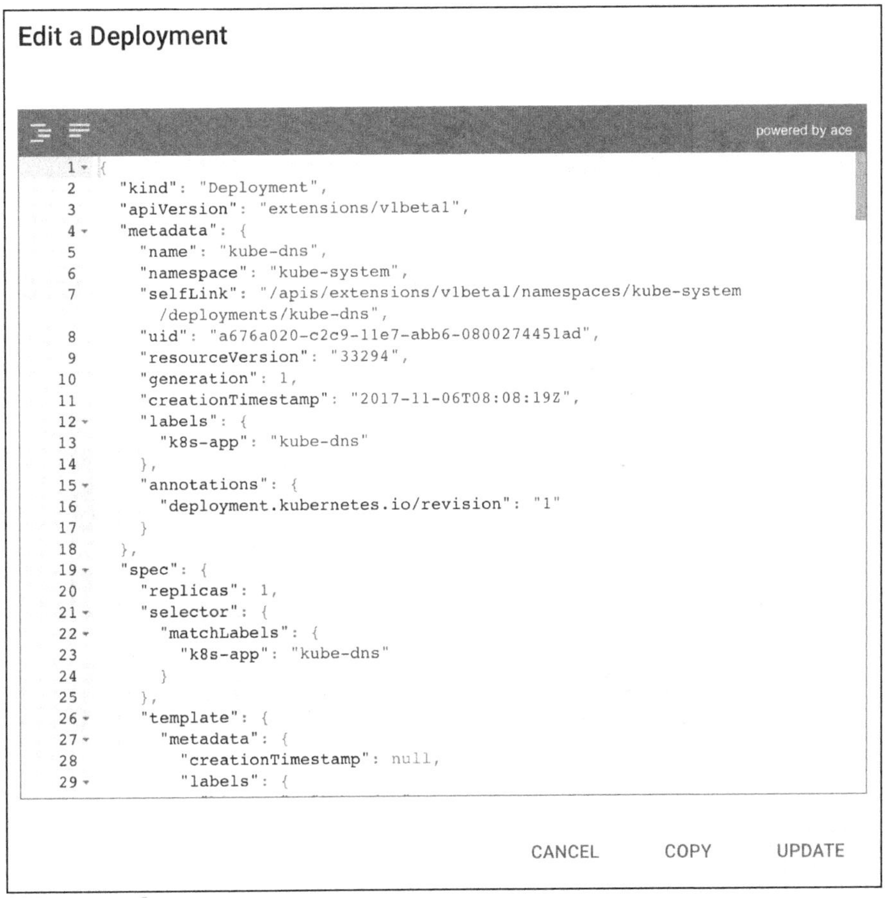

图 13-15

13.4.3 查看资源详细信息

单击某个资源实例的名字，可以查看详细信息，其效果与 kubectl describe 一样，如图 13-16 所示。

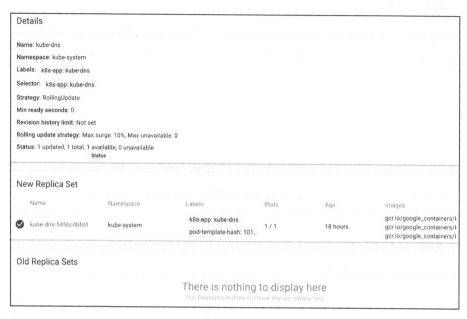

图 13-16

13.4.4 查看 Pod 日志

在 Pod 及其父资源（DaemonSet、ReplicaSet 等）页面单击≡按钮，可以查看 Pod 的日志，其效果与 kubectl logs 一样，如图 13-17 所示。

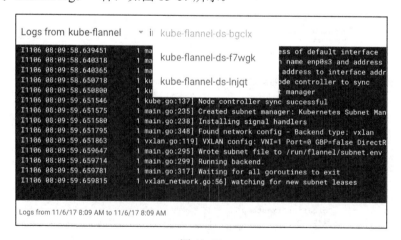

图 13-17

Kubernetes Dashboard 界面设计友好，自解释性强，可以看作 GUI 版的 kubectl，更多功能留给大家自己探索。

13.5 小结

本章介绍了 Kubernetes Dashboard 的安装和使用方法。Dashboard 能完成日常管理的大部分工作，可以作为命令行工具 kubectl 的有益补充。

第 14 章 Kubernetes 集群监控

创建 Kubernetes 集群并部署容器化应用只是第一步。一旦集群运行起来，我们需要确保集群一起都是正常的，所有必要组件就位并各司其职，有足够的资源满足应用的需求。Kubernetes 是一个复杂系统，运维团队需要有一套工具帮助他们获知集群的实时状态，并为故障排查提供及时和准确的数据支持。

本章重点讨论 Kubernetes 常用的监控方案，下一章会讨论日志管理。

14.1 Weave Scope

Weave Scope 是 Docker 和 Kubernetes 可视化监控工具。Scope 提供了自上而下的集群基础设施和应用的完整视图，用户可以轻松对分布式的容器化应用进行实时监控和问题诊断。

14.1.1 安装 Scope

安装 Scope 的方法很简单，执行如下命令：

```
kubectl apply --namespace kube-system -f
"https://cloud.weave.works/k8s/scope.yaml?k8s-version=$(kubectl version | base64 | tr -d '\n')"
```

部署成功后，有如图 14-1 所示的相关组件。

```
ubuntu@k8s-master:~$
ubuntu@k8s-master:~$ kubectl get --namespace=kube-system daemonset weave-scope-agent
NAME                DESIRED   CURRENT   READY     UP-TO-DATE   AVAILABLE   NODE SELE
weave-scope-agent   3         3         3         3            3           <none>
ubuntu@k8s-master:~$
ubuntu@k8s-master:~$ kubectl get --namespace=kube-system deployment weave-scope-app
NAME              DESIRED   CURRENT   UP-TO-DATE   AVAILABLE   AGE
weave-scope-app   1         1         1            1           18h
ubuntu@k8s-master:~$
ubuntu@k8s-master:~$ kubectl get --namespace=kube-system service weave-scope-app
NAME              TYPE       CLUSTER-IP       EXTERNAL-IP   PORT(S)        AGE
weave-scope-app   NodePort   10.106.149.68    <none>        80:30693/TCP   18h
ubuntu@k8s-master:~$
ubuntu@k8s-master:~$ kubectl get --namespace=kube-system pod | grep weave
weave-scope-agent-765g2                        1/1       Running   0   18h
weave-scope-agent-nchjr                        1/1       Running   0   18h
weave-scope-agent-wnzn4                        1/1       Running   0   18h
weave-scope-app-567cfdb6d5-kk2cg                1/1       Running   0   18h
ubuntu@k8s-master:~$
```

图 14-1

（1）DaemonSet weave-scope-agent，集群每个节点上都会运行的 scope agent 程序，负责收集数据。

（2）Deployment weave-scope-app，scope 应用，从 agent 获取数据，通过 Web UI 展示并与用户交互。

（3）Service weave-scope-app，默认是 ClusterIP 类型，为了方便，已通过 kubectl edit 修改为 NodePort。

14.1.2 使用 Scope

浏览器访问 http://192.168.56.106:30693/，Scope 默认显示当前所有的 Controller（Deployment、DaemonSet 等），如图 14-2 所示。

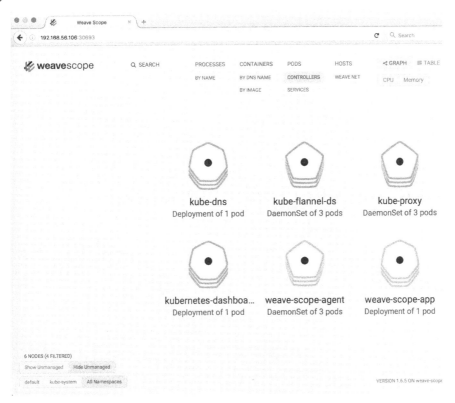

图 14-2

1. 拓扑结构

Scope 会自动构建应用和集群的逻辑拓扑，比如单击顶部 PODS，会显示所有 Pod 以及 Pod 之间的依赖关系，如图 14-3 所示。

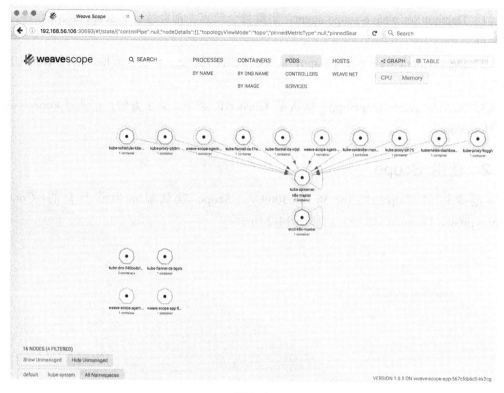

图 14-3

单击 HOSTS，会显示各个节点之间的关系，如图 14-4 所示。

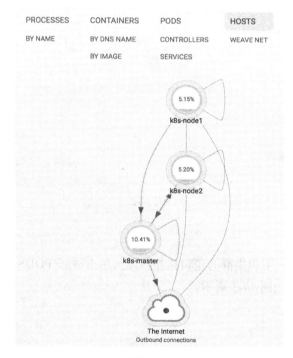

图 14-4

2. 实时资源监控

可以在 Scope 中查看资源的 CPU 和内存使用情况，如图 14-5 所示。
支持的资源有 Host、Pod 和 Container，如图 14-6、图 14-7 所示。

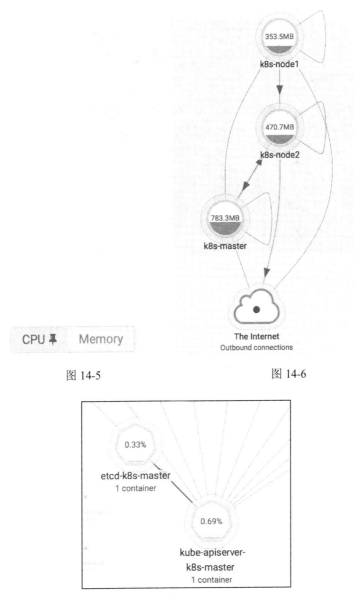

图 14-5　　　　　　　　　　　图 14-6

图 14-7

3. 在线操作

Scope 还提供了便捷的在线操作功能，比如选中某个 Host，单击 >_ 按钮可以直接在浏览器中打开节点的命令行终端，如图 14-8 所示。

图 14-8

单击 Deployment 的 + 可以执行 Scale Up 操作，如图 14-9 所示。

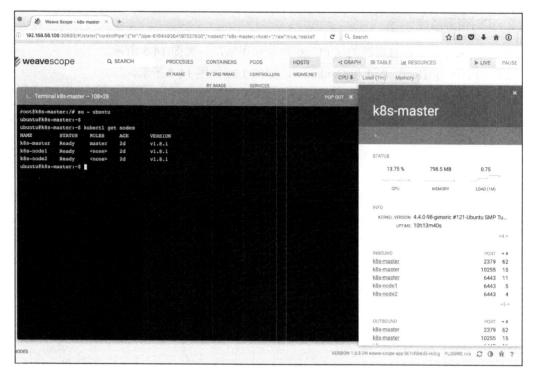

图 14-9

查看 Pod 的日志，如图 14-10 所示。

第 14 章 Kubernetes 集群监控

图 14-10

可以查看 attach、restart、stop 容器，以及直接在 Scope 中排查问题，如图 14-11 所示。

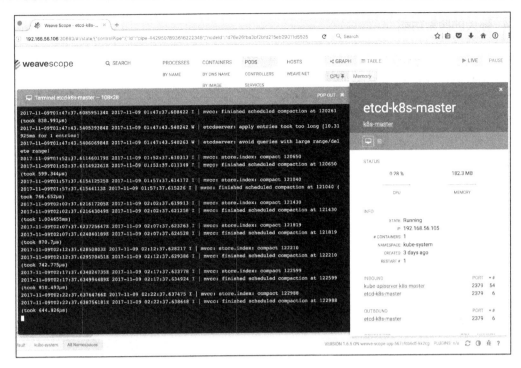

图 14-11

4. 强大的搜索功能

Scope 支持关键字搜索和定位资源，如图 14-12 所示。还可以进行条件搜索，比如查找和定位 MEMORY 大于 100MB 的 Pod，如图 14-13 所示。

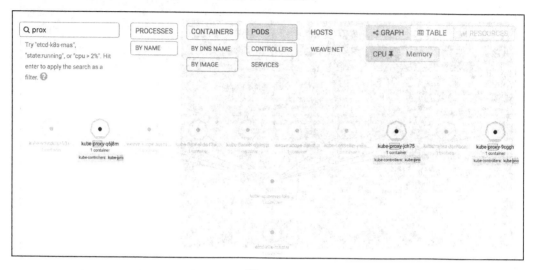

图 14-12

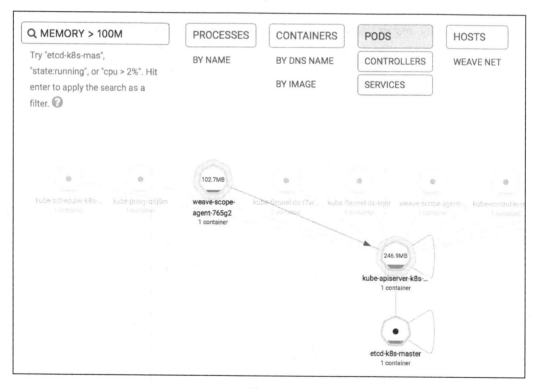

图 14-13

Weave Scope 界面极其友好，操作简洁流畅，更多功能留给大家去探索。

14.2 Heapster

Heapster 是 Kubernetes 原生的集群监控方案。Heapster 以 Pod 的形式运行，它会自动发现集群节点，从节点上的 Kubelet 获取监控数据。Kubelet 则是从节点上的 cAdvisor 收集数据。

Heapster 将数据按照 Pod 进行分组，将它们存储到预先配置的 backend 并进行可视化展示。Heapster 当前支持的 backend 有 InfluxDB（通过 Grafana 展示）、Google Cloud Monitoring 等。Heapster 的整体架构如图 14-14 所示。

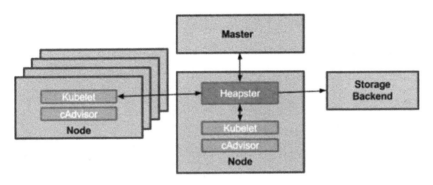

图 14-14

下面我们将实践由 Heapster、InfluxDB 和 Grafana 组成的监控方案。Kubelet 和 cAdvisor 是 Kubernetes 的自带组件，无须额外部署。

14.2.1 部署

Heapster 本身是一个 Kubernetes 应用，部署方法很简单，运行如下命令：

```
git clone https://github.com/kubernetes/heapster.git
kubectl apply -f heapster/deploy/kube-config/influxdb/
kubectl apply -f heapster/deploy/kube-config/rbac/heapster-rbac.yaml
```

Heapster 相关资源如图 14-15 所示。

```
ubuntu@k8s-master:~$
ubuntu@k8s-master:~$ kubectl get --namespace=kube-system deployment | grep -e heapster -e monitor
heapster                  1    1    1    1    8m
monitoring-grafana        1    1    1    1    8m
monitoring-influxdb       1    1    1    1    8m
ubuntu@k8s-master:~$
ubuntu@k8s-master:~$ kubectl get --namespace=kube-system service | grep -e heapster -e monitor
heapster              ClusterIP   10.108.228.4    <none>   80/TCP         8m
monitoring-grafana    NodePort    10.111.8.115    <none>   80:32314/TCP   8m
monitoring-influxdb   ClusterIP   10.99.44.147    <none>   8086/TCP       8m
ubuntu@k8s-master:~$
ubuntu@k8s-master:~$ kubectl get --namespace=kube-system pod -o wide | grep -e heapster -e monitor
heapster-5d67855584-5hdqn                 1/1   Running   0   8m   10.244.1.18   k8s-node2
monitoring-grafana-5bccc9f786-5bbkw       1/1   Running   0   8m   10.244.2.13   k8s-node1
monitoring-influxdb-85cb4985d4-t2b26      1/1   Running   0   8m   10.244.2.14   k8s-node1
ubuntu@k8s-master:~$
```

图 14-15

为了便于访问，已通过 kubectl edit 将 Service monitoring-grafana 的类型修改为 NodePort。

14.2.2 使用

浏览器打开 Grafana 的 Web UI：http://192.168.56.105:32314/。

Heapster 已经预先配置好了 Grafana 的 DataSource 和 Dashboard，如图 14-16 所示。

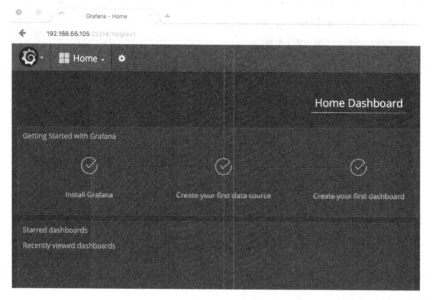

图 14-16

单击左上角的 Home 菜单，可以看到预定义的 Dashboard Cluster 和 Pods，如图 14-17 所示。

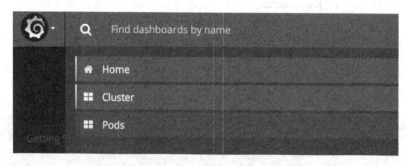

图 14-17

单击 Cluster，可以查看集群中节点的 CPU、内存、网络和磁盘的使用情况，如图 14-18 所示。

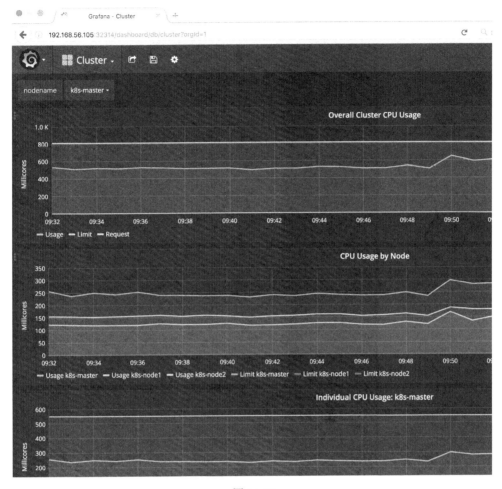

图 14-18

在左上角可以切换查看不同节点的数据,如图 14-19 所示。

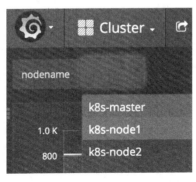

图 14-19

切换到 Pods Dashboard,可以查看 Pod 的监控数据,包括单个 Pod 的 CPU、内存、网络和磁盘使用情况,如图 14-20 所示。

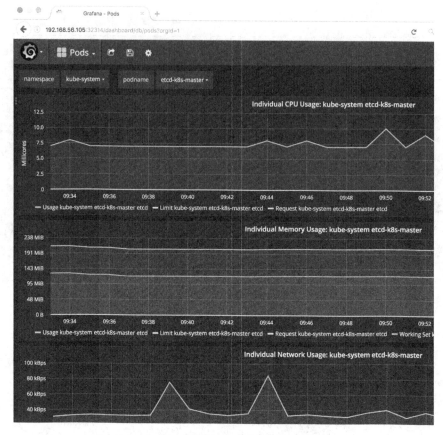

图 14-20

在左上角可以切换到不同 Namespace 的 Pod，如图 14-21 所示。

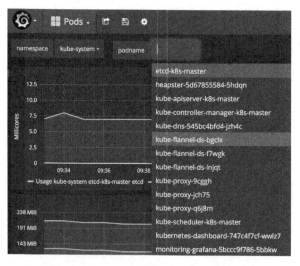

图 14-21

Heapster 预定义的 Dashboard 很直观，也很简单。如有必要，可以在 Grafana 中定义自己的 Dashboard，满足特定的业务需求。

14.3 Prometheus Operator

前面我们介绍了 Kubernetes 的两种监控方案，即 Weave Scope 和 Heapster，它们主要的监控对象是 Node 和 Pod。这些数据对 Kubernetes 运维人员是必需的，但还不够。我们通常还希望监控集群本身的运行状态，比如 Kubernetes 的 API Server、Scheduler、Controller Manager 等管理组件是否正常工作以及负荷是否过大等。

本节我们将学习监控方案 Prometheus Operator，它能回答上面这些问题。

Prometheus Operator 是 CoreOS 开发的基于 Prometheus 的 Kubernetes 监控方案，也可能是目前功能最全面的开源方案。我们先通过截图了解一下它能干什么。

Prometheus Operator 通过 Grafana 展示监控数据，预定义了一系列的 Dashboard，如图 14-22 所示。

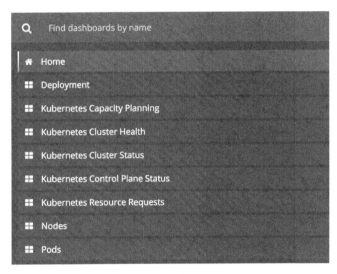

图 14-22

- Kubernetes 集群的整体健康状态如图 14-23 所示。

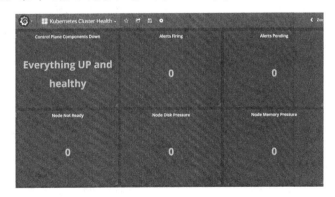

图 14-23

- 整个集群的资源使用情况如图 14-24、图 14-25 所示。

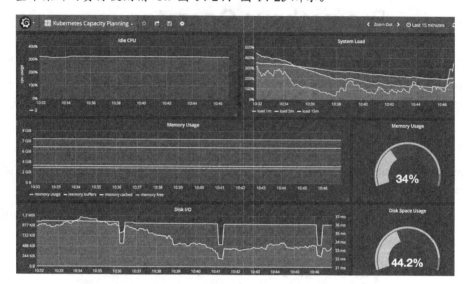

图 14-24

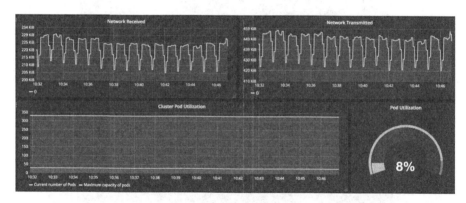

图 14-25

- Kubernetes 各个管理组件的状态如图 14-26、图 14-27 所示。

图 14-26

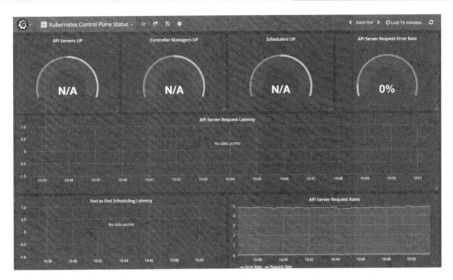

图 14-27

- 节点的资源使用情况如图 14-28 所示。

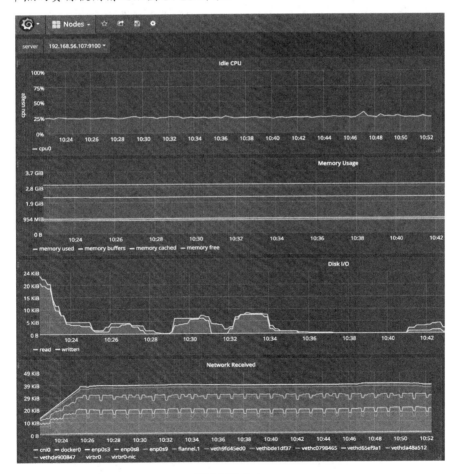

图 14-28（a）

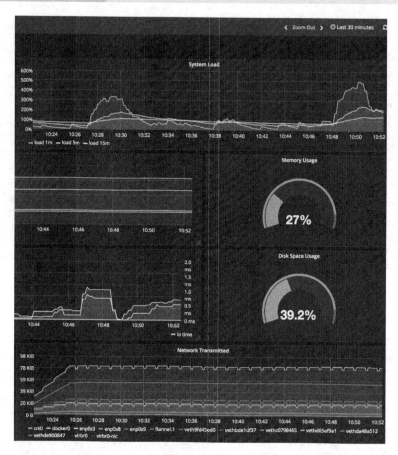

图 14-28（b）

- Deployment 的运行状态如图 14-29 所示。

图 14-29

- Pod 的运行状态如图 14-30 所示。

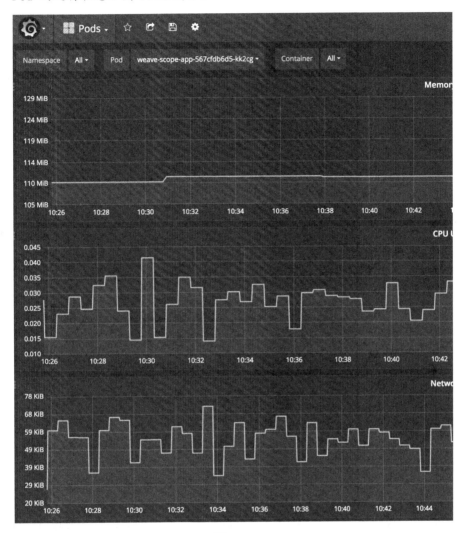

图 14-30

这些 Dashboard 展示了从集群到 Pod 的运行状况，能够帮助用户更好地运维 Kubernetes，而且 Prometheus Operator 迭代非常快，相信会继续开发出更多更好的功能，所以值得我们花些时间学习和实践。

14.3.1　Prometheus 架构

因为 Prometheus Operator 是基于 Prometheus 的,所以我们需要先了解一下 Prometheus。

Prometheus 是一个非常优秀的监控工具。准确地说应该是监控方案。Prometheus 提供了数据搜集、存储、处理、可视化和告警一套完整的解决方案。Prometheus 的架构如图 14-31 所示。

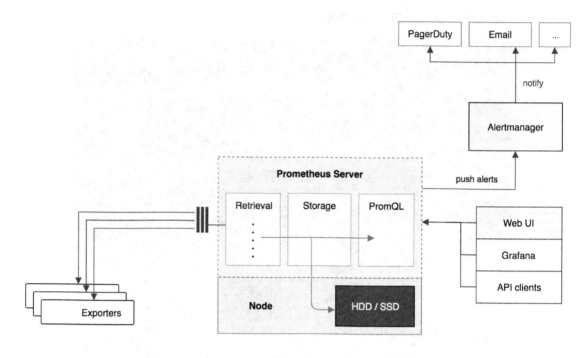

图 14-31

官网上的原始架构图比上面这张要复杂一些，为了避免注意力分散，这里只保留了最重要的组件。

1. Prometheus Server

Prometheus Server 负责从 Exporter 拉取和存储监控数据，并提供一套灵活的查询语言（PromQL）供用户使用。

2. Exporter

Exporter 负责收集目标对象（host、container 等）的性能数据，并通过 HTTP 接口供 Prometheus Server 获取。

3. 可视化组件

监控数据的可视化展现对于监控方案至关重要。以前 Prometheus 自己开发了一套工具，不过后来废弃了，因为开源社区出现了更为优秀的产品 Grafana。Grafana 能够与 Prometheus 无缝集成，提供完美的数据展示能力。

4. Alertmanager

用户可以定义基于监控数据的告警规则，规则会触发告警。一旦 Alermanager 收到告警，就会通过预定义的方式发出告警通知，支持的方式包括 Email、PagerDuty、Webhook 等。

14.3.2 Prometheus Operator 架构

Prometheus Operator 的目标是尽可能简化在 Kubernetes 中部署和维护 Prometheus 的工作。其架构如图 14-32 所示。

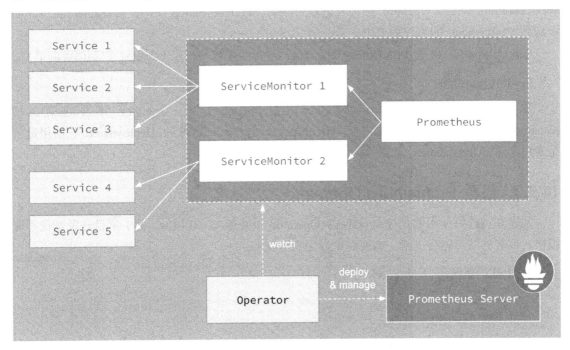

图 14-32

图 14-32 中的每一个对象都是 Kubernetes 中运行的资源。

1. Operator

Operator 即 Prometheus Operator，在 Kubernetes 中以 Deployment 运行。其职责是部署和管理 Prometheus Server，根据 ServiceMonitor 动态更新 Prometheus Server 的监控对象。

2. Prometheus Server

Prometheus Server 会作为 Kubernetes 应用部署到集群中。为了更好地在 Kubernetes 中管理 Prometheus，CoreOS 的开发人员专门定义了一个命名为 Prometheus 类型的 Kubernetes 定制化资源。我们可以把 Prometheus 看作一种特殊的 Deployment，它的用途就是专门部署 Prometheus Server。

3. Service

这里的 Service 就是 Cluster 中的 Service 资源，也是 Prometheus 要监控的对象，在 Prometheus 中叫作 Target。每个监控对象都有一个对应的 Service。比如要监控 Kubernetes Scheduler，就得有一个与 Scheduler 对应的 Service。当然，Kubernetes 集群默认是没有这个 Service 的，Prometheus Operator 会负责创建。

4. ServiceMonitor

Operator 能够动态更新 Prometheus 的 Target 列表，ServiceMonitor 就是 Target 的抽象。比如想监控 Kubernetes Scheduler，用户可以创建一个与 Scheduler Service 相映射的 ServiceMonitor 对象。Operator 则会发现这个新的 ServiceMonitor，并将 Scheduler 的 Target 添加到 Prometheus 的监控列表中。

ServiceMonitor 也是 Prometheus Operator 专门开发的一种 Kubernetes 定制化资源类型。

5. Alertmanager

除了 Prometheus 和 ServiceMonitor，Alertmanager 是 Operator 开发的第三种 Kubernetes 定制化资源。我们可以把 Alertmanager 看作一种特殊的 Deployment，它的用途就是专门部署 Alertmanager 组件。

14.3.3 部署 Prometheus Operator

笔者在实践时使用的是 Prometheus Operator 最新版本 v0.14.0。由于项目开发迭代速度很快，部署方法可能会更新，必要时请参考官方文档。

1. 下载最新源码

```
git clone https://github.com/coreos/prometheus-operator.git
cd prometheus-operator
```

为方便管理，创建一个单独的 Namespace monitoring，Prometheus Operator 相关的组件都会部署到这个 Namespace。

```
kubectl create namespace monitoring
```

2. 安装 Prometheus Operator Deployment

```
helm install --name prometheus-operator --set rbacEnable=true --namespace=monitoring helm/prometheus-operator
```

Prometheus Operator 所有的组件都打包成 Helm Chart，安装部署非常方便，如图 14-33 所示。如果对 Helm 不熟悉，可以参考前面相关的章节。

```
ubuntu@k8s-master:~$
ubuntu@k8s-master:~$ kubectl get --namespace=monitoring deployment prometheus-operator
NAME                                       DESIRED   CURRENT   UP-TO-DATE   AVAILABLE
prometheus-operator-prometheus-operator    1         1         1            1
ubuntu@k8s-master:~$
```

图 14-33

3. 安装 Prometheus、Alertmanager 和 Grafana

```
helm install --name prometheus --set serviceMonitorsSelector.app=prometheus --set ruleSelector.app=prometheus --namespace=monitoring helm/prometheus
  helm install --name alertmanager --namespace=monitoring helm/alertmanager
```

```
helm install --name grafana --namespace=monitoring helm/grafana
```
可以通过 kubectl get prometheus 查看 Prometheus 类型的资源，如图 14-34 所示。

```
ubuntu@k8s-master:~$
ubuntu@k8s-master:~$ kubectl get --namespace=monitoring prometheus
NAME         AGE
prometheus   1d
ubuntu@k8s-master:~$
ubuntu@k8s-master:~$ kubectl get --namespace=monitoring pod prometheus-prometheus-0
NAME                      READY    STATUS    RESTARTS   AGE
prometheus-prometheus-0   2/2      Running   0          1d
ubuntu@k8s-master:~$
ubuntu@k8s-master:~$ kubectl get --namespace=monitoring service prometheus-prometheus
NAME                    TYPE       CLUSTER-IP      EXTERNAL-IP   PORT(S)         AGE
prometheus-prometheus   NodePort   10.96.207.169   <none>        9090:30413/TCP  1d
ubuntu@k8s-master:~$
```

图 14-34

为了方便访问 Prometheus Server，这里已经将 Service 类型通过 kubectl edit 改为 NodePort。

同样可以查看 Alertmanager 和 Grafana 的相关资源，如图 14-35、图 14-36 所示。

```
ubuntu@k8s-master:~$
ubuntu@k8s-master:~$ kubectl get --namespace=monitoring alertmanager
NAME           AGE
alertmanager   2d
ubuntu@k8s-master:~$
ubuntu@k8s-master:~$ kubectl get --namespace=monitoring pod alertmanager-alertmanager-0
NAME                          READY   STATUS    RESTARTS   AGE
alertmanager-alertmanager-0   2/2     Running   0          2d
ubuntu@k8s-master:~$
ubuntu@k8s-master:~$ kubectl get --namespace=monitoring service alertmanager-alertmanager
NAME                        TYPE       CLUSTER-IP      EXTERNAL-IP   PORT(S)         AGE
alertmanager-alertmanager   NodePort   10.103.53.199   <none>        9093:32758/TCP  2d
ubuntu@k8s-master:~$
```

图 14-35

```
ubuntu@k8s-master:~$
ubuntu@k8s-master:~$ kubectl get --namespace=monitoring deployment grafana-grafana
NAME              DESIRED   CURRENT   UP-TO-DATE   AVAILABLE   AGE
grafana-grafana   1         1         1            1           5h
ubuntu@k8s-master:~$
ubuntu@k8s-master:~$ kubectl get --namespace=monitoring pod grafana-grafana-5fdf676c68-v7dlb
NAME                               READY   STATUS    RESTARTS   AGE
grafana-grafana-5fdf676c68-v7dlb   2/2     Running   0          5h
ubuntu@k8s-master:~$
ubuntu@k8s-master:~$ kubectl get --namespace=monitoring service grafana-grafana
NAME              TYPE       CLUSTER-IP       EXTERNAL-IP   PORT(S)        AGE
grafana-grafana   NodePort   10.109.107.156   <none>        80:32342/TCP   5h
ubuntu@k8s-master:~$
```

图 14-36

Service 类型也都已经改为 NodePort。

4. 安装 kube-prometheus

kube-prometheus 是一个 Helm Chart，打包了监控 Kubernetes 需要的所有 Exporter 和 ServiceMonitor。

```
helm install --name kube-prometheus --namespace=monitoring helm/kube-prometheus
```

每个 Exporter 会对应一个 Service，为 Pormetheus 提供 Kubernetes 集群的各类监控数据，如图 14-37 所示。

图 14-37

每个 Service 对应一个 ServiceMonitor，组成 Pormetheus 的 Target 列表，如图 14-38 所示。

图 14-38

与 Prometheus Operator 相关的所有 Pod 如图 14-39 所示。

图 14-39

我们注意到有些 Exporter 没有运行 Pod，这是因为像 API Server、Scheduler、Kubelet 等 Kubernetes 内部组件原生就支持 Prometheus，只需要定义 Service 就能直接从预定义端口获取监控数据。

通过浏览器打开 Pormetheus 的 Web UI（http://192.168.56.105:30413/targets），如图 14-40 所示。

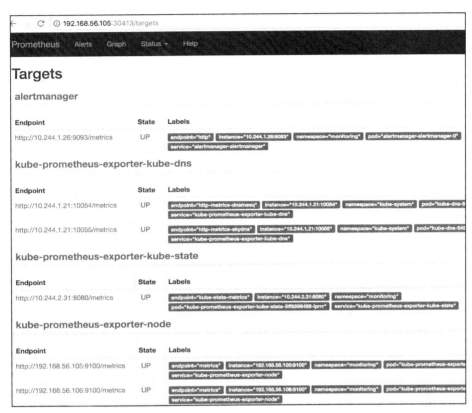

图 14-40

可以看到所有 Target 的状态都是 UP。

5. 安装 Alert 规则

Prometheus Operator 提供了默认的 Alertmanager 告警规则，通过如下命令安装。

```
sed -ie 's/role: prometheus-rulefiles/app: prometheus/g' contrib/kube-prometheus/manifests/prometheus/prometheus-k8s-rules.yaml
    sed -ie 's/prometheus: k8s/prometheus: prometheus/g' contrib/kube-prometheus/manifests/prometheus/prometheus-k8s-rules.yaml
    sed -ie 's/job=\"kube-controller-manager/job=\"kube-prometheus-exporter-kube-controller-manager/g' contrib/kube-prometheus/manifests/prometheus/prometheus-k8s-rules.yaml
    sed -ie 's/job=\"apiserver/job=\"kube-prometheus-exporter-kube-api/g' contrib/kube-prometheus/manifests/prometheus/prometheus-k8s-rules.yaml
    sed -ie 's/job=\"kube-scheduler/job=\"kube-prometheus-exporter-kube-scheduler/g'
```

```
contrib/kube-prometheus/manifests/prometheus/prometheus-k8s-rules.yaml
    sed -ie 's/job=\"node-exporter/job=\"kube-prometheus-exporter-node/g'
contrib/kube-prometheus/manifests/prometheus/prometheus-k8s-rules.yaml
    kubectl apply -n monitoring -f
contrib/kube-prometheus/manifests/prometheus/prometheus-k8s-rules.yaml
```

6. 安装 Grafana Dashboard

Prometheus Operator 定义了显示监控数据的默认 Dashboard，通过如下命令安装。

```
    sed -ie 's/grafana-dashboards-0/grafana-grafana/g'
contrib/kube-prometheus/manifests/grafana/grafana-dashboards.yaml
    sed -ie 's/prometheus-k8s.monitoring/prometheus-prometheus.monitoring/g'
contrib/kube-prometheus/manifests/grafana/grafana-dashboards.yaml
    kubectl apply -n monitoring -f
contrib/kube-prometheus/manifests/grafana/grafana-dashboards.yaml
```

打开 Grafana 的 Web UI（http://192.168.56.105:32342/），如图 14-41 所示。

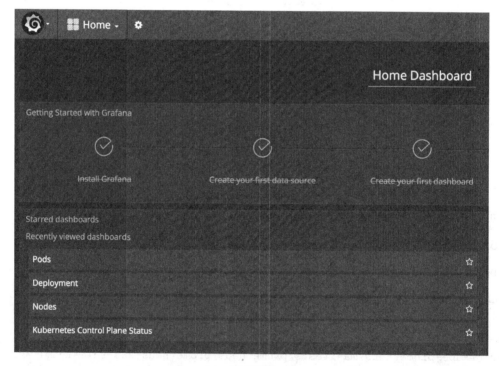

图 14-41

Grafana 的 DataSource 和 Dashboard 已自动配置，单击 Home 就可以使用我们在最开始讨论过的那些 Dashboard 了，如图 14-42 所示。

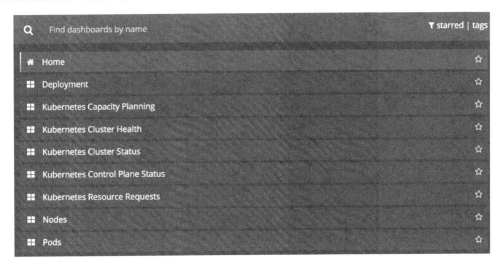

图 14-42

14.4 小结

本章我们实践了三种 Kubernetes 监控方案。

（1）Weave Scope 可以展示集群和应用的完整视图。其出色的交互性让用户能够轻松对容器化应用进行实时监控和问题诊断。

（2）Heapster 是 Kubernetes 原生的集群监控方案。预定义的 Dashboard 能够从 Cluster 和 Pods 两个层次监控 Kubernetes。

（3）Prometheus Operator 可能是目前功能最全面的 Kubernetes 开源监控方案。除了能够监控 Node 和 Pod，还支持集群的各种管理组件，比如 API Server、Scheduler、Controller Manager 等。

Kubernetes 监控是一个快速发展的领域，随着 Kubernetes 的普及，一定会涌现出更多的优秀方案。

第 15 章
Kubernetes 集群日志管理

Kubernetes 开发了一个 Elasticsearch 附加组件来实现集群的日志管理。这是 Elasticsearch、Fluentd 和 Kibana 的组合。Elasticsearch 是一个搜索引擎，负责存储日志并提供查询接口；Fluentd 负责从 Kubernetes 搜集日志并发送给 Elasticsearch；Kibana 提供了一个 Web GUI，用户可以浏览和搜索存储在 Elasticsearch 中的日志，如图 15-1 所示。

图 15-1

15.1 部署

Elasticsearch 附加组件本身会作为 Kubernetes 的应用在集群里运行，其 YAML 配置文件可从 https://github.com/kubernetes/kubernetes/tree/master/cluster/addons/fluentd-elasticsearch 获取，如图 15-2 所示。

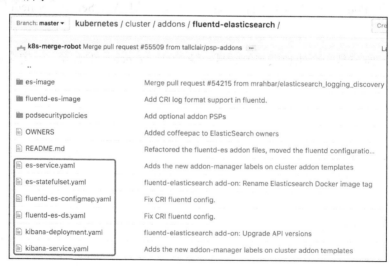

图 15-2

可将这些 YAML 文件下载到本地目录，比如 addons，通过 kubectl apply -f addons/ 部署，如图 15-3 所示。

```
ubuntu@k8s-master:~$
ubuntu@k8s-master:~$ kubectl apply -f addons/
service "elasticsearch-logging" created
serviceaccount "elasticsearch-logging" created
clusterrole "elasticsearch-logging" created
clusterrolebinding "elasticsearch-logging" created
statefulset "elasticsearch-logging" created
configmap "fluentd-es-config-v0.1.1" created
serviceaccount "fluentd-es" created
clusterrole "fluentd-es" created
clusterrolebinding "fluentd-es" created
daemonset "fluentd-es-v2.0.2" created
deployment "kibana-logging" created
service "kibana-logging" created
ubuntu@k8s-master:~$
```

图 15-3

这里有一点需要注意：后面我们会通过 NodePort 访问 Kibana，需要注释掉 kibana-deployment.yaml 中的环境变量 SERVER_BASEPATH，否则无法访问，如图 15-4 所示。

```
spec:
  containers:
  - name: kibana-logging
    image: docker.elastic.co/kibana/kibana:5.6.2
    resources:
      # need more cpu upon initialization, therefore burstable class
      limits:
        cpu: 1000m
      requests:
        cpu: 100m
    env:
      - name: ELASTICSEARCH_URL
        value: http://elasticsearch-logging:9200
     # - name: SERVER_BASEPATH
     #   value: /api/v1/proxy/namespaces/kube-system/services/kibana-logging
      - name: XPACK_MONITORING_ENABLED
        value: "false"
      - name: XPACK_SECURITY_ENABLED
        value: "false"
    ports:
    - containerPort: 5601
      name: ui
      protocol: TCP
```

图 15-4

所有的资源都部署在 kube-system Namespace 里，如图 15-5 所示。

```
ubuntu@k8s-master:~$ kubectl get --namespace=kube-system daemonset fluentd-es-v2.0.2
NAME                DESIRED   CURRENT   READY   UP-TO-DATE   AVAILABLE   NODE SELECTOR   AGE
fluentd-es-v2.0.2   2         2         2       2            2           <none>          9m
ubuntu@k8s-master:~$
ubuntu@k8s-master:~$ kubectl get --namespace=kube-system pod -l "k8s-app=fluentd-es"
NAME                      READY   STATUS    RESTARTS   AGE
fluentd-es-v2.0.2-2hjp4   1/1     Running   0          9m
fluentd-es-v2.0.2-m4gq7   1/1     Running   0          9m
ubuntu@k8s-master:~$
```

图 15-5

DaemonSet fluentd-es 从每个节点收集日志，然后发送给 Elasticsearch，如图 15-6 所示。

```
ubuntu@k8s-master:~$ kubectl get --namespace=kube-system statefulset elasticsearch-logging
NAME                    DESIRED   CURRENT   AGE
elasticsearch-logging   2         2         13m
ubuntu@k8s-master:~$
ubuntu@k8s-master:~$ kubectl get --namespace=kube-system pod -l "k8s-app=elasticsearch-logging"
NAME                      READY     STATUS    RESTARTS   AGE
elasticsearch-logging-0   1/1       Running   0          13m
elasticsearch-logging-1   1/1       Running   0          13m
ubuntu@k8s-master:~$
ubuntu@k8s-master:~$ kubectl get --namespace=kube-system service elasticsearch-logging
NAME                    TYPE       CLUSTER-IP     EXTERNAL-IP   PORT(S)          AGE
elasticsearch-logging   NodePort   10.103.27.59   <none>        9200:32607/TCP   14m
ubuntu@k8s-master:~$
```

图 15-6

Elasticsearch 以 StatefulSet 资源运行，并通过 Service elasticsearch-logging 对外提供接口。这里已经将 Service 的类型通过 kubectl edit 修改为 NodePort。

可通过 http://192.168.56.106:32607/ 验证 Elasticsearch 已正常工作，如图 15-7 所示。

```
← → C  ① 192.168.56.106:32607

{
  "name" : "elasticsearch-logging-1",
  "cluster_name" : "kubernetes-logging",
  "cluster_uuid" : "wRgkHHpNRGCtyQoSvpAIjA",
  "version" : {
    "number" : "5.6.2",
    "build_hash" : "57e20f3",
    "build_date" : "2017-09-23T13:16:45.703Z",
    "build_snapshot" : false,
    "lucene_version" : "6.6.1"
  },
  "tagline" : "You Know, for Search"
}
```

图 15-7

Kibana 以 Deployment 资源运行，用户可通过 Service kibana-logging 访问其 Web GUI。这里已经将 Service 的类型修改为 NodePort，如图 15-8 所示。

```
ubuntu@k8s-master:~$
ubuntu@k8s-master:~$ kubectl get --namespace=kube-system deployment kibana-logging
NAME             DESIRED   CURRENT   UP-TO-DATE   AVAILABLE   AGE
kibana-logging   1         1         1            1           21m
ubuntu@k8s-master:~$
ubuntu@k8s-master:~$ kubectl get --namespace=kube-system pod -l "k8s-app=kibana-logging"
NAME                              READY     STATUS    RESTARTS   AGE
kibana-logging-7879c88776-sfhnv   1/1       Running   0          21m
ubuntu@k8s-master:~$
ubuntu@k8s-master:~$ kubectl get --namespace=kube-system service kibana-logging
NAME             TYPE       CLUSTER-IP      EXTERNAL-IP   PORT(S)          AGE
kibana-logging   NodePort   10.103.43.140   <none>        5601:30319/TCP   21m
ubuntu@k8s-master:~$
```

图 15-8

通过 http://192.168.56.106:30319/ 访问 Kibana，如图 15-9 所示。

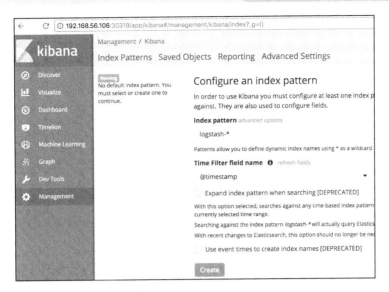

图 15-9

Kibana 会显示 Index Pattern 创建页面。直接单击 Create，Kibana 会自动完成后续配置，如图 15-10 所示。

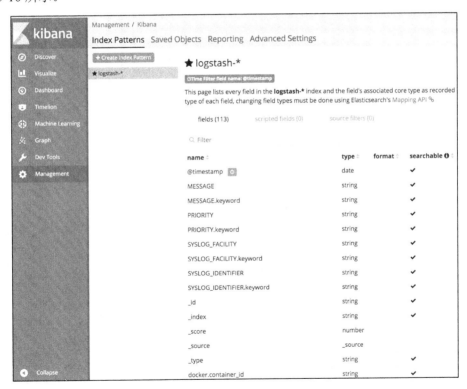

图 15-10

这时，单击左上角的 Discover 就可以查看和检索 Kubernetes 日志了，如图 15-11 所示。

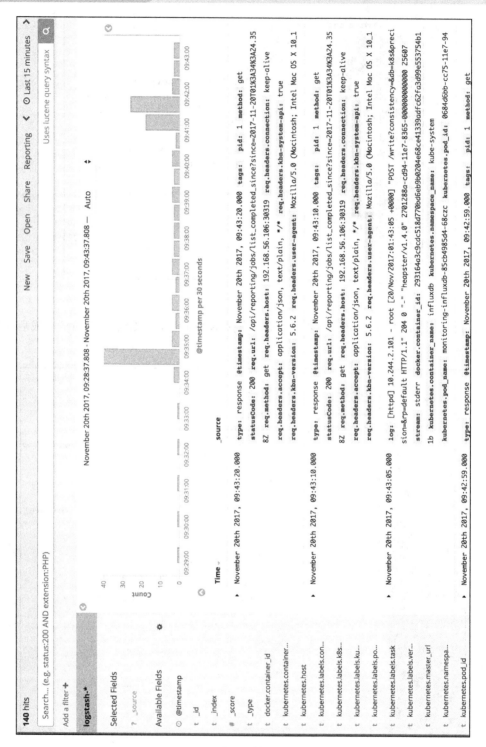

图 15-11

Kubernetes 日志管理系统已经就绪,用户可以根据需要创建自己的 Dashboard,具体方法可参考 Kibana 官方文档。